War Bound from Stockton

U.S. Navy Ships from California's Central Valley, In Harm's Way in Pacific Waters in World War II

Cdr. David D. Bruhn, U.S. Navy (Retired)

At Stockton, California, during World War II, nearly three dozen small, wooden-hulled Navy ships stood out from builders' yards alongside its shipping channel, destined for combat duty in the Pacific. The thirty-two salvage ships, rescue tugs, minesweepers, patrol craft sweepers, and net laying ships owed their existence in part to President Franklin D. Roosevelt. A former Assistant Secretary of the Navy and yachtsman, he recognized upon America's entry into the war that existing yards churning out steel ships as fast as possible could still not meet the Navy's burgeoning requirements. Accordingly, he directed work to large numbers of underutilized boat and yacht building yards; resulting in thousands of wooden ships and craft, including the famous PT boats and Higgins boats, joining the Fleet. Twenty-three Stockton ships collectively earned thirty-eight battle stars, and one, a Presidential Unit Citation. The latter award is the highest a unit may receive for heroism, and the equivalent of the Silver Star Medal for an individual. Following the war, the ships' crews, mostly reservists ("hostilities only" personnel in Royal Navy parlance), returned home to their civilian lives, and the Navy quickly disposed of almost all of the wooden vessels. Americans wanted a peace dividend, and the nation no longer needed the service of the stalwart sailors, nor their valiant little ships. One hundred fifty-nine photographs, maps, and diagrams; appendices; and an index to full names, places, and subjects add value to this work.

To the men and women who toiled in Stockton boat- and shipyards, doing their best for America's war effort, and the officers and men who sailed the sturdy little ships produced there in harm's way, in distant waters far from home.

War Bound from Stockton

U.S. Navy Ships from California's Central Valley, In Harm's Way in Pacific Waters in World War II

Cdr. David D. Bruhn, U.S. Navy (Retired)

HERITAGE BOOKS
2023

HERITAGE BOOKS
AN IMPRINT OF HERITAGE BOOKS, INC.

Books, CDs, and more—Worldwide

For our listing of thousands of titles see our website
at
www.HeritageBooks.com

Published 2023 by
HERITAGE BOOKS, INC.
Publishing Division
5810 Ruatan Street
Berwyn Heights, Md. 20740

International Standard Book Number
Paperbound: 978-0-7884-2912-5

Heritage Books by Cdr. David D. Bruhn, USN (Retired)

Battle Stars for the "Cactus Navy":
America's Fishing Vessels and Yachts in World War II

Enemy Waters:
Royal Navy, Royal Canadian Navy, Royal Norwegian Navy, U.S. Navy, and Other Allied Mine Forces Battling the Germans and Italians in World War II
Cdr. David D. Bruhn, USN (Retired)
and Lt. Cdr. Rob Hoole, RN (Retired)

Eyes of the Fleet:
The U.S. Navy's Seaplane Tenders and Patrol Aircraft in World War II

Gators Offshore and Upriver:
The U.S. Navy's Amphibious Ships and Underwater Demolition Teams, and Royal Australian Navy Clearance Divers in Vietnam

Guns Up, Depth Charges Readied:
U.S. Navy, Commonwealth, and Other Allied Escort Ships Shepherding Convoys, and Battling German and Italian Air and Naval Forces in the Mediterranean in World War II

Guns Up:
Naval Action in the Yellow Sea off Korea, 1950–1953

Home Waters:
Royal Navy, Royal Canadian Navy, and U.S. Navy Mine Forces Battling U-Boats in World War I
Cdr. David D. Bruhn, USN (Retired)
and Lt. Cdr. Rob Hoole, RN (Retired)

Ingram's Fourth Fleet:
U.S. and Royal Navy Operations Against German Runners, Raiders, and Submarines in the South Atlantic in World War II

Intercept:
The U.S. Navy's Intelligence-Gathering Ships ("Cold War Spy Fleet") 1961–1969, 1985–1989

Kissing Cousins:
U.S. Navy Wooden Minesweepers and Variants (YMS, PCS, AGS) and USN and Royal Australian Navy Bomb and Mine Disposal Personnel in the Pacific in World War II, 1944–1945

MacArthur and Halsey's "Pacific Island Hoppers":
The Forgotten Fleet of World War II

Nightraiders:
U.S. Navy, Royal Navy, Royal Australian Navy, and Royal Netherlands Navy Mine Forces Battling the Japanese in the Pacific in World War II
Cdr. David D. Bruhn, USN (Retired)
and Lt. Cdr. Rob Hoole, RN (Retired)

On the Gunline:
U.S. Navy and Royal Australian Navy Warships off Vietnam, 1965–1973
Cdr. David D. Bruhn, USN (Retired)
and STGCS Richard S. Mathews, USN (Retired)

Queenstown Bound:
U.S. Navy Destroyers Combating German U-boats
in European Waters in World War I

Ready to Haul, Ready to Fight:
U.S. Navy, Royal Australian Navy, and British Merchant
Navy Cargo Ships in the Pacific in World War II

Salvation from the Sky:
U.S. Navy, Royal Australian Air Force, and Royal New Zealand Air
Force Heroic Air-Sea Rescue in the Pacific in World War II
Cdr. David D. Bruhn, USN (Retired)
and Stephen Ekholm

Send Some King's Ships:
U.S. Navy, Royal Naval Patrol Service, and Royal Canadian Navy Ships
Combating German U-boats off North America's Eastern Seaboard,
and RNPS and South African Naval Forces Vessels in
African Waters as well, 1942–1945
Cdr. David D. Bruhn, USN (Retired)
and Lt. Cdr. Rob Hoole, RN (Retired)

Stride Out

Support for the Fleet:
U.S. Navy and Royal Australian Navy Service
Force Ships That Served in Vietnam, 1965–1973

Toe the Mark

Turn into the Wind:
Volume I: US Navy and Royal Navy Light Fleet Aircraft Carriers
in World War II, and Contributions of the British Pacific Fleet

Turn into the Wind:
Volume II: US Navy, Royal Navy, Royal Australian Navy, and
Royal Canadian Navy Light Fleet Aircraft Carriers in the
Korean War and through End of Service, 1950–1982

War Bound from Stockton: U.S. Navy Ships from California's
Central Valley, In Harm's Way in Pacific Waters in World War II

We Are Sinking, Send Help!:
The U.S. Navy's Tugs and Salvage Ships in the African,
European, and Mediterranean Theaters in World War II

Wooden Ships and Iron Men:
The U.S. Navy's Coastal and Motor Minesweepers, 1941–1953

Wooden Ships and Iron Men:
The U.S. Navy's Coastal and Inshore Minesweepers, and
the Minecraft that Served in Vietnam, 1953–1976

Wooden Ships and Iron Men:
The U.S. Navy's Ocean Minesweepers, 1953–1994

Contents

Photos and Illustrations

Maps and Diagrams

Acknowledgements

These acknowledgements recognize individuals who provided much assistance with the book. They also explain the significance of the cover art by Richard DeRosset, and the context provided in the forewords penned by George Duddy, David Rajkovich, and Commodore Hector Donohue AM RAN (Ret.). Additionally, some larger Stockton, California, maritime heritage relating to the book will be found here.

Brilliant, prolific, and extraordinarily accomplished maritime artist Richard DeRosset captured in the cover art, a vivid image of the patrol craft sweeper USS *PCS-1404* in action against Japanese forces ashore on Aguijan Island in World War II. The subject of Chapter 1, this small, unglamorous ship, a product of Colberg Boat Works in Stockton, was close off a very small, little-known island in the lower Marianas, carrying out her duty, and contributing to the Allied effort in the Pacific to defeat Japanese aggression, and bring a close to the war in the Pacific.

Richard, who resides and creates his fine maritime and aviation art in San Diego, is an extremely accomplished and prolific artist with 1,041 paintings to his credit, not including the many murals he has done. Some examples of his work may be viewed at his artist's website: http://www.derossetpaintings.com.

Photo Acknowledgements-1

Richard DeRosset being honored at the
San Diego Veterans Museum and Memorial Center.
Courtesy of Richard DeRosset

In his foreword, Canadian George Duddy rightly pays tribute to the late naval architect Tim Colton, who in addition to his professional work, amassed a huge database of shipyard construction records. Most of the American and Canadian yards he cites, closed decades ago. The information he compiled is of great importance to researchers and aided the author with this book. Duddy also highlights the vital contributions of small yards, such as those at Stockton, to the overall war effort. As he recounts, George first learned about the agricultural California city after getting lost in San Francisco fog while attempting to drive to Vallejo, and learning that he had gone astray only after coming upon barely discernable road signs announcing Stockton.

George has served as critical reviewer and content editor for a number of my books. In addition to having the keen eye and analytical mind of a professional engineer, he also has much knowledge of maritime subjects. In his retirement, he has developed into an expert on the maritime history of western Canada and the Arctic, but his lineage stretches across the Atlantic to Britain. His father, as a schoolboy, took photographs of surrendered German battleships in the Firth of Forth near the end of World War I; his great grandfather was a pioneering Leith steamship owner and his great great grandfather, as Master in both sail and steam, ended his career as marine superintendent for the Leith, Hull & Hamburg Steam Packet Co. One of his other relatives, Midshipman Percival George, tragically fell to his death from the mast of a Royal Navy ship during the age of sail. His dirk is on display in a museum in South Africa.

Photo Acknowledgements-2

George Duddy in Glacier Bay aboard the MS *Volendam* during an Alaskan cruise in 2019. A contributor and U.S. Military & Naval Vessel Correspondent for Nauticapedia.ca Project, he is sporting an organization ballcap.
Courtesy of George Duddy

George recently authored a very interesting and engrossing book, based on extensive research of primary sources and his considerable personal knowledge of northwestern Canada. Titled *Called by the North: Extraordinary Adventures of the Fur Trade, Shipbuilders, Navigators and Traders in Northwestern Canada and Alaska*, it is available online from the publisher, Heritage Books, at:
https://heritagebooks.com/search?type=product%2Carticle%2Cpage&q=george+duddy

Following the collapse of whaling in Canadian western Arctic waters at the start of the 1900s, vessels facing the perilous voyage around the Alaskan Peninsula came in pursuit of Arctic fur trade. They came initially from the old whaling ports of California and settled locations in Alaska, but after 1914 also from Vancouver on Canada's west coast. The vessels included those owned — or in support of — large fur trading companies and also those of adventurers bent on making their fortunes in the rich trade. Arctic transportation was also provided by expansion of the existing Mackenzie River system from the interior of Canada through the Boreal Forest and Mackenzie Delta. This book provides a fascinating account of the ships, shipbuilders and navigators of these waterways, and how the Arctic fur trade, pioneered by American entrepreneurs, was finally taken over by the Hudson's Bay Company. Expanding eastward, the Company achieved many Arctic "firsts." In 1930 a relay of company vessels successfully made the first west to east transit of the Northwest Passage; and several fur trading posts developed into permanent northern settlements. These, and many other intriguing stories, are enriched through 193 photographs, maps and diagrams; appendices; a bibliography, and an index to full names, places, and subjects, all adding value to this unique work.

The impetus for this book was the continuing efforts, by Stockton resident David Rajkovich, to bring back to life a wooden minesweeper of the same type built in Stockton after the Korean War, and to create a maritime museum to help preserve and honor the rich and relatively unknown maritime history of Stockton. In his foreword for *War Bound from Stockton*, Rajkovich provides an overview of the totality of efforts in World War II put forth by the city and its residents, of which building naval vessels was only a part.

He describes the restoration of the ex-USS *Lucid* (MSO-458). Though not related to the book's subject matter, it is part of Rajkovich's larger efforts to highlight Stockton's maritime significance—with which this book will assist. I first met David in 2011, when I was invited to attend a meeting at his ranch in Stockton, where the mayor, port captain, and other dignitaries were in attendance. My role was to discuss the viability of the minesweeper as a museum ship, having authored a book (*Wooden Ships and Iron Men: The U.S. Navy's Ocean Minesweepers, 1953-1994*) about these type vessels.

Photo Acknowledgements-3

David Rajkovich giving a shipboard tour of USS *Lucid* to local California National Guard troops.
Courtesy of David Rajkovich

Commissioned on 4 May 1955, *Lucid* served the Navy until the latter part of the Vietnam War. She was decommissioned on 23 December 1970, and "disposed of by Navy Sale" on 1 November 1976 to W. Dean Kirkpatrick of San Francisco. William Gardner, a scrapper, acquired the ship in 1986. He towed it to Bradford Island in the Sacramento Delta; removed and sold everything of value from the ship; and used it as a storage building, cutting a hole in the hull on the port side near the waterline for greater access to below deck areas.

In 2005, Gardner's widow donated the ship to Navy veteran Mike Warren, who had served aboard *Lucid* as an engineman in the 1960s. Warren and a group of volunteers, mostly ex-sailors like himself, began the monumental task of restoring *Lucid* by removing tons of junk that had accumulated inside the ship over the years and then repairing the opening in her side. Having done as much as he could, Warren then sought a permanent home for the ex-*Lucid.*

Learning of the ship's availability, Rajkovich, hosted a gathering of interested persons at his home. The ensuing discussion centered on whether or not the post-Korean War minesweeper—constructed of hearty oak keel, stem, transom, frames, and planking—was in sufficient shape and was significant enough to warrant the time, effort, and other resources necessary for her restoration. The consensus was yes, and the ship was moved from Bradford Island to a berth at the former Naval Reserve Center adjacent to Stockton's deep ship channel.

This relocation to the former Reserve Center building has been fortuitous as it is next to the San Joaquin Building Futures Academy, a vocational high school, where directors immediately saw the restoration of the hull as a wonderful teaching project for their students. Their instructors have been guiding and overseeing student work on the ship for the past several years. Concurrently on weekends, volunteers, mostly ex-minesweeper sailors, have added their considerable talents and expertise to the project.

Photo Acknowledgements-4

Ex-USS *Lucid* (MSO-458) under restoration. Her mast was removed to allow passage to Stockton under bridges, and has not yet been replaced. Courtesy of David Rajkovich

Photo Acknowledgements-5

Ship's bridge, ex-USS *Lucid*.
Courtesy of David Rajkovich

David Rajkovich continues to be the visionary and driving force behind the restoration of *Lucid*, and formation of the Stockton Maritime Museum. The ship had been painstakingly restored, space by space, and is nearing readiness for movement to a pristine location on the Stockton waterfront.

Rajkovich is a third-generation farmer born and raised in northern California, with a lifelong interest in military history. Surrounded by men of the Greatest Generation while growing up, Dave always gravitated to these neighbors and friends of his parents after a BBQ or other get togethers, to listen and learn firsthand stories when the talk eventually led to "The War." After college, he began his own farming business, with row crops transitioning to walnuts and apples. The apple business led to the co-founding of Farmington Fresh, a packer, shipper and marketer of fresh fruit to the domestic and international market.

Following the sale of the business, Dave and his wife Karen began devoting time to helping local "Gold Star parents" who had lost a child on military service while in the line of duty, and recently returning young combat veterans from overseas. Monthly get togethers are now hosted on the farm to get these deserving young men and women and their families together to help with the transition back to civilian life. Separately, work on the ship and initiatives associated with the Stockton Maritime Museum continue. The museum will tell the history of Stockton's shipbuilding past; including the craftsmen who built sturdy naval vessels both in wood and steel, among them ocean-going wooden ships, and the iron men who sailed them in harm's way.

Commodore Hector Donohue AM RAN (Rtd.) has generously provided much assistance with previous books and has done so again. He describes in his foreword the Borneo Campaign which culminated in a successful amphibious assault on Balikpapan. Australia suffered more than 2,000 casualties for the entire campaign, during which supporting U.S. Navy minesweepers earned a Presidential Unit Citation for great heroism in clearing deadly mines blocking shoreward movement of assault forces, while under fire from enemy shore batteries.

Donohue began his career in the RAN in 1955 as a seaman officer and subsequently sub-specialized as a clearance diver and torpedo and anti-submarine officer. His service in the RAN included command of the destroyer escort HMAS *Yarra* and the guided missile frigate HMAS *Darwin*. Ashore, he held a number of senior positions in Defence policy and force development prior to retirement in mid-1991.

Commodore Donohue has authored and co-authored many books and articles, including the books: *From Empire Defence to the Long Haul*

Post-war Defence Policy and Its Impact on Naval Force Structure Planning 1945-1955; *United and Undaunted - the First 100 Years*; *Mines, Mining and Mine Countermeasures*, and *Australian Minesweepers at War*. The cover art for and a description of the latter seminal work, follows.

Photo Acknowledgements-6

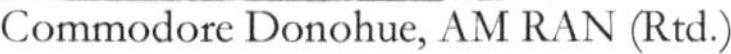

Commodore Donohue, AM RAN (Rtd.)

At the beginning of World War I the emergent RAN was ill-prepared and ill-equipped to counter a mining campaign in home waters. Following German mining in 1917, the RAN quickly requisitioned trawlers as auxiliary minesweepers manned by the RAN Brigade and the mines were ultimately swept.

Both Germany and Japan laid offensive mines in Australian waters during World War II. This stretched the RAN minesweeping effort to the limit, however safe channels were quickly established and the movement of traffic, including vital troopship convoys, was not impeded.

The little known but noteworthy operational minesweeping in China and in the islands post World War II in arduous and difficult conditions was a major feat.

Defensive minefields totalling some 9,000 Australian made mines were laid in 1941-43 by HMAS *Bungaree*, mainly in the Great Barrier Reef, requiring a significant post war clearing operation which resulted in the sinking of one minesweeper after hitting a mine.

This ground breaking book outlines for the first time the significant effort of the RAN's minesweeping forces, which quickly established an expertise in this form of warfare and ensured that the sea lanes remained navigable throughout both wars. The Australian designed and built minesweepers proved their effectiveness in the major sweeping operations during and after World War II.

Finally, many thanks to my long-time editor, Lynn Marie Tosello. She continues to be a stalwart "shipmate" in assisting in bringing to the fore important contributions of unsung small ships and their crews, which would otherwise likely be consigned to the dustbin of history.

Foreword

We must be the great arsenal of democracy. For us this is an emergency as serious as war itself. We must apply ourselves to our task with the same resolution, the same sense of urgency, the same spirit of patriotism and sacrifice as we would show were we at war.

—President Theodore Roosevelt

Photo Foreword-1

Tourist charter boat *Pacific Yellowfin*, ex-U.S. Army freight ship *FS-64* and later junior mine planter *JMP-64*, is one of the original wooden U.S. military vessels still in service. As explained in the text, WWII records for US Army vessels, such as this one, have largely been destroyed and their war stories lost.
Courtesy of John MacFarlane

Unlike David and his local associates who live near and are familiar with Stockton, California, and have much interest and focus in preserving its shipbuilding heritage, my takeaway from this book is quite different. It is that Stockton is a specific example of the much broader and amazing enterprise that permitted the winning of WWII – the massive North American shipbuilding effort.

My association and friendship with David Bruhn began during my research on former U.S. military vessels for an article I was writing in

2016 (*Converted Military Vessels – Western Canada's Maritime War Dividend – Military Vessels Converted for Civilian Use*). I was struck by the fact that he was interested in their fighting lives while I was drawn to their civilian lives as almost 100 former U.S. vessels were purchased at low cost and used in British Columbia's transportation, forest and fishing industries to enormous benefit.

Unlike most readers of this book, I am aware of the existence of Stockton and have been for a long time, but until reading it, remained in a fog about what constitutes that community, much less its wartime history. I entered that fog in December 1967 after passing through one of the many toll gates of the San Francisco Bay Bridge, supposedly on my way to visit a young recently married couple who lived in Vallejo only a short distance from San Francisco. I and a friend were visiting that city and northern California for the very first time as wide-eyed young engineers on holiday from a huge hydro-electric project in frozen northern British Columbia.

It soon became evident that my choice of gate had put me on the wrong route. I was on a busy freeway and unable to see signs soon enough to safely exit and so was trapped into continuing in the direction of the freeway. The lighted overhead signs which were visible indicated a destination. They said "Stockton." When I finally was able to turn around and to arrive in Vallejo it was after 11:00 PM. I stayed only long enough to leave an explanatory note on the darkened residence.

The scourges of the evil Axis powers who plunged the world into WWII could not have been driven back to their roots and destroyed without the shipyards of North America. They included those of the U.S. arsenal of democracy as portrayed by President Roosevelt but also those of Canada. One way that researchers of marine history are aware of the huge expanse of this enterprise is by the gift of shipbuilding tables assembled and published by the late naval architect Tim Colton. These document the products of the hundreds of yards, both small and large, that carried the wars across the seas and finally defeated the scourge of ocean raiders and U-boats which threatened to starve Britain into submission.

Colton's records list over one thousand yards, some large some small, some long-established enterprises and some set up purely for the war effort. While, as to be expected, many were located along the two nations' long coastlines, some were sited in the interior of the continent, great distances from the ocean. The continent's huge system of lakes and rivers tamed into navigable waterways permitted minesweepers to be built on Lake Superior for the RCN and subchasers for the USN in Wisconsin on Lake Michigan. Even landing craft were built in fields in Kansas on the Missouri River. The yards and the community chosen by David Bruhn for his book are examples of these inland enterprises and communities that made such a huge contribution to the Allies' victory in the war.

David's valuable stories tell about what happened to Stockton ships and the men that sailed in them in a comprehensible way. What is almost incomprehensible is that these are a mere fraction of the stories of sailors and shipbuilders that were involved in the total effort. Unlike the examples of David's carefully researched stories, Colton's tables say hardly anything about the lives of the ships listed and those who sailed in them. They do point to the massiveness of the overall effort that is mind boggling. There are thousands of other stories of these ships and their men, some have been and others may still be told, but most are already lost to posterity.

Appendix A documents an impressive number of ships, boats and other vessels built at Stockton during WWII. David has done a yeoman's job in pulling the histories of part of Stockton's products together from the scant information available. Lack of information prevented him from documenting a wider expanse of the vessels. Particularly vexing is ships produced for the U.S. Army. He is well aware that some of their ships such as freight ships and tugs made valuable contributions to the Pacific war, even earning Battle Stars. His research has informed him that records of these vessels have been destroyed. Further, massive amounts of mundane craft such as scows, lighters and patrol boats were probably never documented.

In this book, 32 USN wooden ships built in Stockton are taken to the crucial battles of the Pacific, that David, through his other books has documented so well, and joined with the amazing vast fleets that were assembled to attack the Japanese. Against this drama, David lays out the stories of the Stockton ships, how they were commissioned, prepared, voyaged to war, fought and earned Battle Stars. Although all different in their own way, thousands of ships had similar stories.

George H. S. Duddy
White Rock, British Columbia

Foreword

David Bruhn's latest book, *War Bound from Stockton*, does a great job highlighting how the wartime efforts of Stockton's boat- and shipyards, and the crews of the vessels they produced, contributed to victory over Japan in the Pacific. These yards built many specialized vessels for the Army, Navy, and Coast Guard, amongst them, from the smallest to the largest, were: landing craft, cargo barges, tugboats, balloon barges, barracks barges, net-laying ships, Army coastal freighters and oilers, crash rescue boats, minesweepers, and sections of floating dry docks. Navy YMS minesweepers and variants, patrol craft sweepers, were the most numerous-type built, and saw much action in the Pacific Theater.

A factor in Stockton's founding in 1849 was its unique access, despite being well inland, to water transportation to the Pacific and the world. Our waterfront was a strategic jumping off point for 49ers and supplies heading to the nearby mother lode during the California gold rush. It was just an overnight river boat trip from San Francisco, which was the gathering point and epicenter of gold seekers and supplies from around the world. This tremendous demand for marine transportation kicked off the boat building industry in Stockton, with many of our long-operating local boat yards having been established in the 19th century. These yards became the backbone of the ship building industry that grew to ten yards during the war, with approximately 10,000 personnel working in the yards that operated 24 hours a day, seven days a week.

War Bound from Stockton focuses on the important contributions of the locally-built naval vessels, and there is no need to reiterate in this foreword any of their rich history covered in the book. Instead, readers will be introduced to other contributions made by the Stockton community during World War II, and to current initiatives involving the newly established Stockton Maritime Museum.

STOCKTON ARMY AIR CORPS PILOT TRAINING AND HOSTING OF NAVY SUPPLY SUPPORT FUNCTIONS

America's military planners recognized the importance of our local sea, land and rail assets even before the U.S. entered the war. In 1940, the Army took over our local airport, becoming the Stockton Army Airfield, and the operations of our deep-water port, the Port of Stockton. The Army Air Corps airfield was an advanced pilot training base, turning out a class of pilots every few weeks. To accomplish this feat, the airfield also operated five additional county auxiliary airfields to provide sufficient runways for thousands of trainees to take off from, and land. In fact, many of the pilots of the famous Doolittle's Raiders—who, piloting B-25s launched from the carrier USS *Hornet*, carried out an air strike on Tokyo—were Stockton-trained pilots.

The Navy built the Rough and Ready Naval Supply Depot on farmland along Stockton's Deepwater Channel. Being ninety miles inland from the Pacific coast, it was less susceptible to the threat of Japanese air or naval attack than were coastal installations. It and a sister local depot, the Sharpe Army Depot, became the largest vehicle consignment center in the nation. Vast amounts of wartime supplies were collected and stored, prior to their loading onto cargo ships to supply the Pacific war zones. So many supplies were stockpiled for the planned invasion of Japan, fortunately cancelled, that it took until the 1960s to auction off the unnecessary, surplus halftracks, jeeps, trucks and other gear and supplies to local farmers and dealers.

Photo Foreword-2

U.S. Navy Reserve Fleet Stockton Group, on the San Joaquin River, 1½ miles downstream from downtown Stockton, California. To the right is the Naval Reserve Center on Rough and Ready Island.
U.S. Navy *All Hands* magazine May 1948

COMMUNITY MILITARY AND CIVILIAN SUPPORT

While certainly not unique, Stockton's contribution to the war effort was a great example of an otherwise ordinary city of 52,000 citizens caught up in the middle of the largest wartime effort of the 20th century. Young Stockton men who had never before traveled far from home, were trained at bases throughout the country, prior to being shipped overseas. And local housewives were employed in war industry for the first time, while retired tradesmen returned to employment and lent their expertise to shops and plants which were losing younger workers to the military services.

School-aged children were heavily involved in raising money for war bonds, and Stockton High School students raised the funding required to purchase 275 jeeps that were then shipped from Stockton depots to all branches of the military, through their unique "Jeep a Week" program.

Unfortunately, Stockton played a big role in FDR's Executive Order #9066 when a temporary relocation center was established on the county fairgrounds. At it's peak, over 4,200 American citizens of Japanese ancestry from Stockton and San Joaquin County were detained until more permanent camps were built in Rowher, Arkansas. But in a sign of true patriotism, many of these interned men joined the Army as soon as they turned 18 while in the relocation camps.

FOOD PRODUCTION FOR MILITARY FORCES

Agriculture was Stockton's first and most important industry, and remains so to this day. Plentiful rich farmland adjacent to river transportation combined to make Stockton home to many large fruit packers, processors, grain growers and flour producers. The unique marshlands of the San Joaquin Delta soils led to the invention and development of new tracked-type tractors by the Holt Manufacturing Company. The company founded by Benjamin Holt here in Stockton eventually grew to become the Caterpillar Tractor Company. Along with Harris Manufacturing and Moore Equipment Company, these agricultural equipment manufactures, plus all of their local machine shops, foundries and affiliated suppliers were easily transformed during WWII to become an important part of our nation's "Arsenal of Democracy." The several large fruit and vegetable canning companies located in Stockton worked 24/7 to meet the demands of preserved foods being shipped to our millions of troops and Allies worldwide.

Labor shortages both on the farms and in the canneries found a unique source of labor during the war. During 1943 and 1944, several POW camps were established at our fairgrounds, at Sharpe Depot, and Rough and Ready Naval Supply Depot for prisoners from the European warfront. POWs were formed into work parties, and worked on local farms or canneries six days a week. On Sundays, POWs were allowed to be "checked out" by farmers to help with chores on their local farmsteads. Local Italian and Italian immigrant farming families could pick up an Italian or German POW to chop wood, bale hay or pick crops, and then enjoy a homemade meal and talk with the host family in their native language about "the old country." What a comparison to the stories of local service members who were prisoners of the Japanese or Germans during the same time.

STOCKTON MARITIME MUSEUM

My interest in Stockton's maritime, industrial and military history, along with being the son of a WWII Merchant Mariner led to my becoming one of the founders of the Stockton Maritime Museum. Our museum is currently restoring the last of 65 former Navy, ocean-going wooden-hulled minesweepers built after the Korean War. The ex-USS *Lucid* (MSO-458) is a sister ship of several minesweepers constructed in Stockton in the early 1950s.

Restoration of the ship has been ongoing for some time. Our museum centered around it is being established on a newly acquired site on Stockton's historic downtown waterfront. Here we will tell the story of the city's maritime history, the local firms and citizens associated with the yards or who toiled in them, and the men who gallantly served aboard the resultant ships, and the impact these vessels had in our nations largest and costliest wartime struggle.

Enjoy David's captivating account of Stockton's WWII naval vessels, and contemplate the students, housewives, farmers and war industry workers on Stockton's home front whose hard work and dedication made this part of our "Arsenal of Democracy" possible.

David Rajkovich
president,
Stockton Maritime Museum

Foreword

In *War Bound from Stockton*, David Bruhn describes the wartime activities of ships and craft built for the U.S. Army, Navy, and Coast Guard during World War II in the city of Stockton, California, which is situated on the San Joaquin River, approximately ninety miles inland from San Francisco Bay. Ten shipyards built thirty-two salvage ships, rescue tugs, minesweepers, patrol craft sweepers, and net laying ships. Their exploits during the war in the Pacific are well covered in *War Bound from Stockton.* Chapter 18 relates to the Borneo campaign which is where "Stockton ships" interacted with the Royal Australian Navy (RAN). In this foreword I will provide a summary of the Borneo campaign from an Australian perspective.

Map Foreword-1

Island of Borneo and surrounding areas

Australian forces successfully led the Allied liberation of Borneo, the world's third largest island, from Japanese occupation in the Oboe series of operations. These operations culminated in Oboe Two, the amphibious assault on Balikpapan which was not only the last large scale

Allied operation of World War II but remains Australia's largest ever amphibious assault. As the culmination of the RAN's participation in over 20 South West Pacific Area (SWPA) amphibious landings during World War II, the Balikpapan assault demonstrated the expertise in amphibious operations the RAN had attained.

With rich oil resources and functional aerodromes, the strategic worth of Borneo was debated at the highest levels. Borneo's position at the base of the South China Sea meant that it shared coastal waters with Indochina, Malaya, Sumatra, Java, Celebes and the Philippines. The US Joint Chiefs of Staff proposed an invasion of Borneo to secure oil and rubber supplies, and interdict Japanese communications with Southeast Asia.

Initial plans called for six Oboe operations, however, as the Allied offensives progressed closer to Japan, Oboe Three, Four and Five were cancelled. The remaining three amphibious landings were code-named: Oboe One, the invasion of Tarakan Island; Oboe Six, the invasion of north Borneo at Labuan and Brunei; and Oboe Two, the invasion of Balikpapan. The sites were selected for the strategic assets and advantages their capture would offer the Allies. Tarakan had an airfield, docking facilities, protected all weather harbourage and relatively good roads. Even without the fleet base option, the liberation of Labuan and Brunei would secure the area's oil and rubber resources. Balikpapan was selected for its oil reserves, two suitable airfields, and deep sheltered harbour.

A prerequisite for the amphibious operations against the Japanese in Borneo was Allied sea and air control. The US Pacific Fleet had decisively defeated Japanese naval forces in the Battle of the Philippine Sea in June 1944 and at the Battle of Leyte Gulf in October 1944. In the meantime, the Allied air forces in the SWPA had advanced along the northern coast of New Guinea and to the Philippines, destroying all enemy forces in their path. An air offensive from northern Australia had destroyed Japanese air power in the Netherlands East Indies and had effectively blockaded Japanese shipping in the area. By early 1945, the Allies had sea and air supremacy across Borneo. At the start of the campaign the Allies estimated that there were 25,000 to 30,000 troops of the Japanese 37th Army and miscellaneous naval units garrisoning Borneo.

The Borneo operations were unique in the respect that serious difficulty was encountered with previously sown allied magnetic mines. The 5th Air Force and RAAF had laid numerous magnetic mines in the shallow waters of Tarakan, Brunei Bay, and Balikpapan, and pre-assault sweeping was conducted for several days in each of these three areas. At Balikpapan where the threat of magnetic mines was especially great, minesweeping continued for two weeks before the landing. Throughout the Borneo operations, the small minesweeping group attached to the 7th Amphibious Force did an outstanding job despite heavy losses—about one ship damaged or lost for every two magnetic mines destroyed.

Photo Foreword-3

Japanese soldiers inspect a mine in shallow water at Balikpapan, Borneo, 17 June 1942. Australian War Memorial photograph

The amphibious assault on Tarakan (Oboe One) commenced, as planned, on 1 May 1945, and despite difficult coastal approaches, extensive minefields and strongly fortified defences, the landing was accomplished with marked success. A heavy concentration of naval and air bombardment prior to the landing, as well as effective naval gunfire support (NGS) to ground forces once ashore effectively neutralised most of the Japanese resistance. Hard fighting by 9th Australian Division troops secured the area.

Photo Foreword-4

Allied invasion craft at Tarakan Island, Borneo, 1 May 1945.
Australian War Memorial photograph OG2674

Photo Foreword-5

Australian soldiers with Allied equipment on the landing beach at Tarakan Island, Borneo, May 1945. The US Navy tank landing ship *LST-530* is discharging stores onto a pontoon pier in the background. The shell holes and shattered palms are evidence of the Allied bombardment.
Australian War Memorial photograph 305124

The landings at Labuan and Brunei (Oboe Six) proceeded to plan on 10 June. After preliminary naval bombardment, hydrographic and mine clearance operations, Australian troops met little Japanese opposition and moved rapidly to their first objectives. NGS helped reduce pockets of resistance on Labuan, while the 9th Australian Division secured much of Brunei and British Borneo. The Australian forces were able to release some Allied prisoners of war as well as provide humanitarian assistance to the Chinese, Malay and Indigenous populations of north Borneo.

The Oboe Two plan required the landing of over 33,000 personnel, their supplies and heavy equipment in the assault. The force included over 21,000 men of the 7th Australian Division, 2,000 Royal Australian Air Force (RAAF) personnel, as well as 2,000 men from United States and Netherlands East Indies units.

The naval forces allocated to Oboe Two under Vice Admiral Daniel E. Barbey, USN, Commander Balikpapan Attack Force, included an Amphibious Task Group, a Cruiser Covering Group, and an Escort Carrier Group. The Amphibious Task Group consisted of over 120 ships, including the RAN Infantry Landing Ships (LSIs), *Manoora* (Flagship of the Transport Unit), *Westralia* and *Kanimbla.* Overall, there were some 98 landing craft and miscellaneous vessels, with a screen of 10 destroyers, five destroyer escorts and the Australian frigate *Gascoyne.* Another frigate, HMAS *Warrego*, operated in a specialised hydrographic unit within this Amphibious Task Group. The Cruiser Covering Group consisted of 10 cruisers and 14 destroyers organised into three separate commands, including HMA Ships *Shropshire* (heavy cruiser), *Hobart*, (light cruiser), *Arunta* (destroyer) and 2 USN destroyers under Commodore Harold B. Farncomb, RAN. The Escort Carrier Group included three carriers with approximately 90 aircraft in total, and six destroyer escorts.

Air support for Oboe Two was supplied by the RAAF, US 13th and 5th Air Forces, and naval air units from the US 3rd and 7th Fleets, and began on 11 June. The RAAF, under Air Vice Marshal William D. Bostock, acted as coordinating agency for all pre-invasion strikes and close support. Altogether, Bostock had 40 squadrons at his disposal for the period, totalling 300 aircraft.

The naval bombardment of Balikpapan commenced on 27 June, with *Shropshire* and *Hobart* firing at Japanese targets along the coast. NGS from all three commands within the Cruiser Covering Group was made available throughout the Oboe Two operations. Over 46,800 rounds were fired by the naval forces in support of the Balikpapan operation, beating all records for ammunition expended in support of a division size landing.

Warrego and the hydrographic unit carried out surveys and placed marker buoys off the landing beaches and surveyed the inner harbour. The mine clearance activities at Balikpapan were some of the most difficult of the war. Sweeping began on 15 June, with 16 minesweepers and a covering force operating in shallow water and uncleared minefields, often under Japanese gunfire. Underwater demolition teams of US Army engineers cleared two gaps through the beach obstacles while under fire. The hydrographic, mine clearance and underwater demolition activities were most successful.

Photo Foreword-6

Last dash to shore, aboard American-manned "Alligators" (LVT amphibious vehicles), during the landing of Australian troops at Balikpapan, Borneo. Smoke from the ruins of enemy positions and burning oil wells is visible in the background.
Australian War Memorial photograph 018812

On 1 July, the first two amphibious waves hit the beaches in 91 amphibious vehicles and despite a choppy sea the ship-to-shore transfer had the troops landing five minutes early at 0855. The Australian LSIs provided parts of the third and subsequent waves. The last of the organised waves, the 17th, landed at 1055. The beaches of Balikpapan were taken with little opposition, and by noon that day, 10,500 troops, 700 vehicles and 1,950 tons of stores had been landed.

Photo Foreword-7

Members of 2/14 Infantry Battalion, disembarking from USS *LCI-999*, and wading ashore at Yellow Beach, Balikpapan, Borneo, 1 July 1945.
Australian War Memorial photograph 110436

Gascoyne escorted a convoy that arrived at Balikpapan on 5 July with supplies essential for the maintenance of land and air forces ashore. The Australian LSIs, having departed as soon as they had unloaded the assault troops, returned with reinforcements from Morotai on 7 July. As the Australian 7th Division advanced inland, they encountered strong pockets of Japanese resistance.

On 15 August 1945, Japan agreed to an unconditional surrender. VP (Victory in the Pacific) Day was celebrated on Borneo, but warily so. Troops were instructed to stay alert. There was no guarantee Japanese garrisons would surrender or had heard the news. No one wanted to be the last killed. In the days following, Australians took the surrender of Japanese across Borneo, and even out to the remote Natuna Islands, in the South China Sea. They also liberated men, women and children of different nationalities from prisoner of war and internment camps.

Although it was not realized at the time, the Balikpapan operation was the last for the 7th Amphibious Force. In the previous two years, Vice Admiral Barbey had planned and conducted 56 amphibious operations in the SWPA, carrying close to a million Allied troops and a million tons of stores from Australia along the north coast of New Guinea, through the Netherlands East Indies, and into the Philippines.

Though questions about their strategic value continued to be asked long after Japan's surrender, the three Oboe operations were successfully and skilfully carried out. The 9th Australian Division carried the brunt of the fighting on land at Tarakan and at Labuan Island, while the 7th Australian Division undertook the Balikpapan operation. All three operations were supported by overwhelmingly powerful Allied air and naval forces. Australian casualties numbered more than 2,000 for the entire campaign. Today we should look back at Borneo 1945 with pride, as it remains a classic example of how Australian strategic interests were successfully pursued through maritime power projection.

Commodore Hector Donohue AM RAN (Rtd)

Preface

Primarily, the SNOWBELL *was a work ship. Her mission was to install and maintain the heavy underwater harbor nets which so effectively protected the ships of our nation and those of our allies during the war. The deck ratings aboard were among the most proficient and hard working in the fleet, necessarily so, since the net business was an exacting one, demanding for success, the close coordination and skills of all hands.*

Her every task was met with a success earned by her hard working crew. In battle, at Okinawa, the SNOWBELL *was fought by the working men who became fighting men with a firm resolution common to all of this nation's fighting men.*

—Ship's history of the net layer USS *Snowbell* (AN-52), dated 29 January 1946.

If a person were asked to name the major ports on the west coast of the United States, and was familiar with the region, they might identify the ports of Seattle, Tacoma, Portland, Oakland, Los Angeles, and Long Beach. If their mental exercise was narrowed to ports in northern California smaller than Oakland, the Ports of Eureka and San Francisco might come to mind, and perhaps others in the San Francisco Bay Area. These might include Port Chicago, as well as the ports of Benicia, Crockett, Richmond, and Redwood City.

Few people would cite the Port of Stockton, which is located deep inland within the agriculturally rich community of California's fertile San Joaquin Valley (see map on following page). The port lies about ninety nautical miles from the Pacific, at the head of a deep ship channel off the San Joaquin River. Exceptions would be those individuals who presently do, or have lived, in Stockton; mariners who have visited there by water; and motorists who have driven the highway that bisects the city. While crossing over an overpass adjacent to the port, it's possible to look down and glimpse the masts of one or more ships below.

The Port of Stockton opened in 1933 as the first inland seaport in California. That year, the first ocean vessel, the cargo ship S.S. *Daisy Gray*, arrived there—a new shipping point for the transportation of agricultural products grown in central California's warm and sunny climate. The City of Stockton is named after Commodore Robert F. Stockton, USN. During the Mexican-American War in the 1840s, while commander of the Pacific Squadron, Stockton's landing force, combined with Army units, was responsible for driving the Mexican forces out of California.[1]

Map Preface-1

San Francisco Bay and adjacent Sacramento-San Joaquin River Delta to the east

Photo Preface-1

Engraving of Commodore Robert F. Stockton, USN
Naval History and Heritage Command photograph NH 44394

Stockton has been a transportation hub since the California Gold Rush days in the mid-1800s, when it was a supply base for the southern mines in the gold country. To provide water access to the city, natural sloughs at Stockton were dredged into a channel allowing usage by river boats as well as local waterborne commerce. In the 1930s, deepening of the channel and dredging of the San Joaquin River leading to it, created a long-sought seaport at Stockton to accommodate deep water steam ships. Further dredging in 1940, and construction of a belt line railroad providing access to all three transcontinental railroads, made the Port of Stockton a strategic site for wartime industries.[2]

WORLD WAR II SHIPBUILDING

In 1940-1941, as the drumbeats of war grew louder, and the likelihood of America's eventual entry into World War II increased, seaport cities on the Atlantic and Pacific coasts vied for government funding to expand urgently needed shipbuilding capacity. Site selection criteria, however, dictated that new yards could not drain off the labor supply from existing yards or overcrowd existing yards' supply channels. The Pacific War had not started and West Coast shipbuilding had not increased to the intensity of that on the East and Gulf coasts closer to the conflicts in Europe. Another factor was the abundant availability of timber in the Pacific Northwest for wooden boat construction. For that reason, cities on the Pacific Coast got a large share of new shipyard construction, which included Stockton, far inland.[3]

A few boat yards in Stockton had been in business since the beginning of the century, building and repairing boats that served local needs, and the city's potential as a maritime center caught the attention of the U.S. Maritime Commission. However, because of the narrow channel at that location, Stockton's several boat- and shipyards operated side-by-side in tight quarters, building small ships and craft, and side-launching them into the restricted waters.[4]

As America entered into the Pacific War, new shipyards quickly popped up. At its zenith, combined with the pre-existing boat yards, Stockton boasted a total of ten boat- and shipyards (identified below):

- Pollock Stockton Shipbuilding
- Colberg Boat Works
- Stephens Brothers
- Hickinbotham Brothers
- Guntert & Zimmerman
- Kyle & Company
- Clyde W. Wood

- Moore Equipment Company
- Nicholson's
- Stockton Steel Fabrications Company[5]

Within a year after the Japanese attack on Pearl Harbor on 7 December 1941, the Inland Shipbuilding Association had formed around the six largest yards, which employed over 3,500 workers. At war's end, over 10,000 men and women were engaged in building ships and craft for the U.S. Army, Navy, and Coast Guard. Appendix A lists the military vessels produced by the yards, with the exception of the small Stockton Steel Fabrications Company, for which the author was unable to locate any records.[6]

The largest employer was Pollock-Stockton Shipbuilding, which leased 50 acres on the north side of the Stockton Channel near Louis Park for the construction of floating dry docks. Pollock's subsequently built rescue boats, invasion barges, supply boats, salvage boats, cargo barges, and seaplane and net tenders, reaching a peak employment of 8,000 in 1944. The first Navy work was a joint contract assigned in 1940 to Colberg Boat Works and Stephens Brothers, for the construction of six-YMS minesweepers.[7]

Colberg's modest capabilities soon grew into one of the largest shipbuilding operations, whose output included minesweepers and sub-chasers as well as salvage, patrol, refueling and lifeboats. The boat works also built salvage, rescue and refrigerator boats for the Royal Navy and, toward the end of the war, repaired and salvaged battle-damaged ships.[8]

Stephens Brothers built minesweepers and rescue boats for the Navy, salvage boats, tugs and rescue craft for the Army, and picket boats for the Coast Guard.[9]

Hickinbotham Brothers, which was established in 1866 as a wagon and carriage shop, evolved to become a hardware supplier and steel jobber, and also entered into a shipbuilding project with Guntert and Zimmerman to build scores of small boats for the Navy. Other small shipbuilding products came from Kyle and Company, Clyde Wood, Moore Company, Nicholson's, and Stockton Steel Fabrications Company.[10]

FIVE TYPES OF SHIPS SUBJECT OF BOOK

Stockton yards built wooden- and steel-hulled ships and craft, most of which served in the Asiatic-Pacific Theater. Only five types of vessels earned battle stars. These were USN salvage ships, rescue tugs, net laying ships, minesweepers, and patrol craft sweepers. Collectively thirty-two wooden-hulled ships, they were the products of three of Stockton's largest and most capable yards—Pollock Stockton Shipbuilding, Colberg Boat Works, and Stephens Brothers.

All are listed in the following table including those of their sisters that did not earn any battle stars. Those that earned one or more stars are denoted with ★ symbols. The yard minesweeper *YMS-95* also earned a Presidential Unit Citation—the highest award for heroism that may be bestowed on a military unit—one equivalent to the Silver Star Medal for an individual. An outlier vessel type built by Stephens Brothers was a civilian motor boat *Sea Breeze*, that was later converted to the Navy ambulance boat known as *YH-4*, she also earned a battle star and is included in the table. Little is known about her war service, other than that her star was awarded for duty at Okinawa.

Types of Stockton-built Ships (and One Boat) that Earned Battle Stars in the Asiatic-Pacific Theater in World War II

Salvage Ships (3)	Minesweepers (12)	Net Laying Ships (10)
Anchor (ARS-13)★	*YMS-94*★	*Lancewood* (AN 48)★
Protector (ARS-14)	*YMS-95*★★ **PUC**	*Papaya* (AN 49)★★
Extractor (ARS-15)	*YMS 96*★★	*Cinnamon* (AN-50)★
Rescue Tugs (4)	*YMS-97*★★★	*Silverbell* (AN 51)★
ATR-50	*YMS 98*	*Snowbell* (AN 52)★
ATR-51★★	*YMS-99*★	*Spicewood* (AN 53)★
ATR-52★	*YMS-383*★	*Manchineel* (AN-54)
ATR-53★	*YMS-384*	*Torchwood* (AN-55)
Ambulance Boat (1)	*YMS-385*	*Winterberry* (AN 56)★
YH-4 (ex-*Sea Breeze*)★	*YMS-386*★	*Viburnum* (AN 57)
	YMS-387★	
	YMS-388★★★★	
	Patrol Craft Sweepers (3)	
	PCS-1402★★★	
	PCS-1403★★★★	
	PCS-1404★★★	

A photograph of *YH-4*, and a representative one of each type ship—rescue and salvage ships, rescue tugs, net laying ships, minesweepers, and patrol craft sweepers—follow.

Wooden-hulled Ambulance Boat *YH-4*
Length 41 feet; Displacement 24 tons; Beam 11' 9"; Draft 3.7'

Photo Preface-2

Ambulance Boat *YH-4* (ex-civilian motor boat *Sea Breeze*), circa mid-1943.
Naval History and Heritage Command photograph #NH 100123

Wooden-hulled *Anchor*-class Rescue and Salvage Ship
Length 183' 3"; Displacement 1,615 tons; Beam 37'; Draft 14' 8"

Photo Preface-3

Rescue and salvage ship USS *Anchor* (ARS-13) under way near the Colberg Boat Works, Stockton, California, probably during acceptance trials in October 1943.
U.S. Navy Bureau of Ships photograph

Wooden-hulled *ATR-1* class Rescue Tug
Length 165' 6"; Displacement 1,315 tons; Beam 33' 4"; Draft 15' 6"

Photo Preface-4

Rescue tug USS *ATR-52* under way in San Francisco Bay, 14 August 1944.
National Archives photograph #19-N-71102

Wooden-hulled *Ailanthus*-class Net Laying Ship
Length 194' 6"; Displacement; 1,400 tons; Beam 37'; Draft 13' 6"

Photo Preface-5

Net laying ship USS *Lancewood* (AN-48) under way, location and date unknown.
Naval History and Heritage Command photograph #NH 73440

YMS-1 class Auxiliary Motor Minesweeper
Length 136'; Displacement 270 tons; Beam 24' 6"; Draft 8'

Photo Preface-6

Yard minesweeper USS *YMS-97* in San Francisco Bay, 1945-1946.
Naval History and Heritage Command #NH 81376

PCS-1376 class Patrol Craft Sweeper
Length 136'; Displacement 338 tons; Beam 24' 6"; Draft 8' 7"

Photo Preface-7

Patrol craft sweeper USS *PCS-1403*, circa 1953.
Philippine Navy photograph courtesy of NavSource

FDR SPURS CONSTRUCTION OF LARGE NUMBERS OF WOODEN-HULLED SHIPS

Thousands of wooden ships and craft served in the U.S. Navy during World War II. This armada, which some fleet sailors derisively termed the "splinter fleet," was comprised initially of former civilian and commercially-owned yachts and fishing vessels—one being the previously referenced motor boat *Sea Breeze*, converted to Ambulance Boat *YH-4*—and subsequently newly constructed wooden-hulled ships purposely designed for Navy tasks. These included the above-described vessels built in Stockton and scores of other yards; as well as the well-known and glamorous PT boats; desperately needed submarine chasers, particularly early in the war; and other lesser-known types of vessels, including coastal transports (APcs). (For more information on the converted yachts and fishing vessels, see my book, *Battle Stars for the "Cactus Navy"*.)[11]

The construction of wooden ships in Stockton yards was only a small part of a much-larger, massive effort, involving boat- and shipyards along America's east, west, and gulf coasts, and Great Lakes. The production of so many wooden vessels was due in part to President Franklin D. Roosevelt, earlier an avid yachtsman. While Assistant Secretary of the Navy during World War I, he recognized that America's large shipyards were taxed in building warships and merchant vessels but that countless smaller boat yards were not directly involved in the nation's war effort, and were available to lend much-needed assistance.[12]

During that war, Roosevelt knew that some of these yards had been building armed motor launches and patrol boats for Britain, France, and Russia, and that suitable small warships, made of wood to get around the steel shortage, could be mass-produced quickly and cheaply in these boat yards as a stopgap measure until larger ships became available. The result of an initiative made by him was a group of 110-foot wooden submarine chasers for the USN.[13]

Following America's entry into World War II, President Roosevelt was aware that Navy requirements for ships and craft greatly exceeded the ability of yards building steel-hulled ships. While these yards were working at full capacity, many of the nation's boat and yacht building yards were not, as had been the case in WWI. Further, there was a ready supply of lumber particularly in the Pacific Northwest that could be quickly acquired and diverted to shipbuilding projects. Once again, Roosevelt helped direct Maritime Commission-contracted work to very capable, smaller yards.[14]

BATTLE HONOURS AND BATTLE STARS

There are references to Royal Australian Navy ships earning battle honours for action described in this book, so an explanation of the significance of these honours is appropriate. The ships of the Royal Navy and the other Commonwealth navies, with successful war service, earn "Battle Honours." Because His/Her Majesty's warships (whether RN, or units of other Commonwealth nations) do not carry Army regimental colours, battle honours are instead displayed on a battle honour board. Traditionally teak, this solid wooden board is mounted on the ship's superstructure, carved with the ship's badge and scrolls naming the ship and the associated honours. The board is either completely unpainted, or with the lettering painted gold. It displays the ships' motto. To pay tribute to past ships of the same name, their honours are also displayed on the board as well.

Photo Preface-8

Battle Honours board of the Royal Australian Navy infantry landing ship HMAS *Manoora*, depicting the ship's battle honours and badge. Australian War Memorial photograph 118843

Battle Honours (which date back to the year 1588, when 'ARMADA 1588' was authorized for the first honor ever) are awarded for six types of action:

- Fleet or Squadron Actions
- Single-ship or Boat Service Actions
- Major Bombardments
- Combined Operations
- Campaign Awards
- Area Awards[15]

Photo Preface-9

Ribbons board on the bridge wing of the battleship USS *New Jersey* (BB-62). Her Asiatic-Pacific campaign ribbon, with one silver star and four bronze stars (denoting nine battle stars), is the one located in the center, third row from the top. Courtesy of John Werda

The United States had a similar method of honoring naval vessels and their crews based on campaign ribbons and "Battle Stars." Naval personnel serving in the Pacific in World War II warranted sporting on their uniform blouses an unadorned Asiatic-Pacific campaign ribbon. Those whose ships earned one or more battle stars were authorized to affix the requisite number of stars to their campaign ribbon. Ships similarly proudly exhibited their Asiatic-Pacific campaign ribbon or those of other war theaters, displayed with other type ribbons earned, on their deckhouses and appropriately adorned with their battle stars.

The *Navy and Marine Corps Awards Manual*, 1953, specified that U.S. Navy ships and units had to meet one of the following criteria to be considered to have participated in combat operations (and thereby earn a battle star):

- Engaged the enemy
- Participated in ground action
- Engaged in aerial flights over enemy territory
- Took part in shore bombardment, minesweeping, or amphibious assault
- Engaged in or launched commando-type raids or other operations behind enemy lines
- Engaged in redeployment under enemy fire
- Engaged in blockade of Korean waters (Korean War)
- Operated as part of carrier task groups from which offensive air strikes were launched
- Were part of mobile logistic support forces in combat areas[16]

In large opposed amphibious island assaults in the Pacific Theater, which is the subject of *War Bound from Stockton*, many U.S. Navy ships and craft earned battle stars. In the book's text, for the sake of brevity, battle stars and their qualifying dates, are limited and only cited for the particular ships or groups of ships, highlighted in the narrative about an event or the role of those particular ships within a larger operation.

ISLAND CHAINS, ATOLLS AND THEIR ISLANDS

Within the vast Pacific Theater, fighting took place in widely scattered archipelagoes (island groups) which often included, in addition to individual islands, larger atolls, themselves made up of small islands and islets. This type of land form can make nomenclature difficult to understand. For example, as shown on the map, within the Marshall Islands are many atolls as well as individual islands. Kwajalein is one such atoll. Its 97 islands and islets (totaling a land area of 6.33 square miles) surround one of the largest lagoons in the world, with an area of 839 square miles.

It is common for the largest island of an atoll to have the same name as the atoll. Kwajalein Island, with an area of about 1.2 square miles, is the southernmost and largest of the islands in Kwajalein Atoll. To the immediate southwest of Kwajalein Atoll is independent Lib Island, which is quite small at only 0.36 square miles.

Map Preface-2

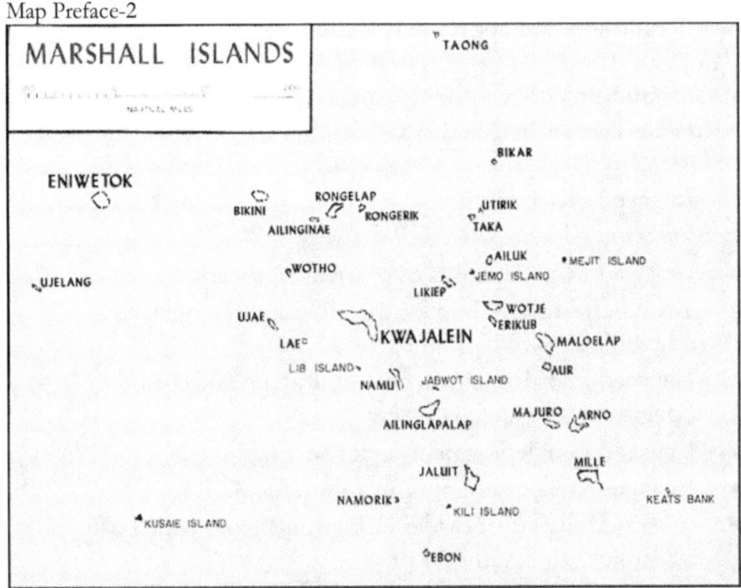

Marshall Islands

A DEARTH OF AVAILABLE INFORMATION

> *The Army's fleet was always considerably below "secondary" to Infantry, Armor, and Artillery. Support services simply do not get attention. It could be argued that the Army considered its ships as floating trucks. The Army's peculiar indifference to its seagoing past might also be a somewhat spiteful reaction to losing these assets to the Navy in the consolidation into a Department of Defense (DoD) in 1949-1950. The Army, despite its meticulous attention to its history, involved itself in the large-scale destruction of its fleet's records – particularly ship logs – during the period of losing the function during the creation of the DoD and demise of the separate cabinet-level War and Navy Departments.*
>
> —From the article "the ghost fleet" by Ian E. Watts.[17]

Readers curious about the collective efforts of the Stockton boat and shipyards, after quickly reviewing Appendix A, may have noted that Hickinbotham Brothers, Guntert & Zimmerman, Kyle & Company, Stephens Brothers, and Clyde W. Wood built vessels for the U.S. Army. There is no coverage of these vessels in this book because of a paucity of information about them, for a variety of reasons as highlighted in the above quoted material.

One of my previous books, titled *MacArthur and Halsey's "Pacific Island Hoppers,"* takes up the subject of Army freight-supply ships, large tugs, coastal tankers, and harbor tugs that served in the Southwest Pacific Theater. None referenced are Stockton-built ships although some of these types were built there. However, the description of the book's contents, provides an overview of Army ship activities in the Pacific War. It and the cover art for the book follow. Because Army ships, including coastal freighters, coastal tankers, and small tugs, were built at Stockton, it would have been desirable to have included coverage of them in this book. Unfortunately, as noted, lack of or destruction of records made them unavailable for reference, although their significant contribution is noted.

MacArthur and Halsey's "Pacific Island Hoppers" The Forgotten Fleet of World War II

At the commencement of World War II, the Navy and the Army—woefully lacking small ships able to ply shallow, reef-infested South and Southwest Pacific waters, which were necessary to support island ground combat—initially acquired whatever was available in ports, harbors, and backwaters to meet their needs. These vessels included schooners, ancient ferry boats, luggers, fishing trawlers, tuna boats, tugs, launches, lighters, surf boats, ketches, yachts, and yawls. The services took whatever craft they could get—some barely seaworthy—as the urgency of need did not permit discrimination in what was purchased or chartered. Gen. Douglas MacArthur, needing his own Navy to support leapfrog operations up the New Guinea coast, found his vessels in Australia and New Zealand, and the Navy its small ships and craft in America. These "Pacific island hoppers" were later supplemented with other small vessels newly constructed in American boat and shipyards. Among them were sixty Navy wooden-hulled 103-foot small coastal transports, hundreds of Army freight-supply ships and large tugs, and lesser numbers of coastal tankers and harbor tugs. The Army ships—most of steel construction, a few of wood—were manned by Coast Guard, Merchant Marine, or Army crews. The island hoppers worked mostly with amphibious forces, but also supported PT boat squadrons, and as "maids of all duties" engaged in a variety of operations. Periodic combat with Japanese planes off the New Guinea coast and in the Solomon Islands transitioned to frequent battles with conventional and kamikaze aircraft and suicide Q-boats during the Philippine Islands Campaign. Significant numbers of the island hoppers earned battle stars, and crewmen awards for valor including the Navy Cross, the Silver Star and the Bronze Star medals. Following the war, the Navy acquired some of the Army ships; many served in the Korean War and a few in Vietnam. Three of the former freight-supply ships were employed for intelligence gathering; the most famous, USS *Pueblo*, was captured by North Korea. Others led interesting careers under civilian ownership; one was run aground while engaged in drug smuggling in the Caribbean, and another served as a "radio pirate" off England, broadcasting BBC-banned rock and roll music over the airwaves in 1966.

NAVY SHIPS' WAR DIARIES: LACK THEREOF AND POOR QUALITY

Following America's entry into World War II, Adm. Ernest King, USN, commander in chief, U.S. Fleet, and chief of Naval Operations, mandated that U.S. Navy ships and shore commands begin maintaining war diaries as of April 1942. Despite this decree, many small ships did not do so for a variety of reasons, but likely primarily attributable to a lack of manpower. Small ships have many of the same functions and responsibilities of larger ships, so with their small crews, fewer personnel to accomplish proportionally greater workloads, this requirement tended to fall by the wayside.

It was not uncommon aboard small ships for the commanding officer to stand bridge watches, in addition to carrying out many other important responsibilities, leaving little time to put pen to paper. Additionally, being junior officers, some captains of small ships may have believed that their senior counterparts aboard larger ships, generally privy to more information, could more succinctly describe day-to-day operations, including those against enemy forces.

It is important to note that all ships kept deck logs. However, these mostly only contained required information such as ship movements including course and speed changes, weather observations, disciplinary action related to crew infractions, etc. Aboard small ships, the logs submitted to the Navy as required, contain mostly handwritten and not typewritten entries common for larger ships, and sloppy writing or the availability of only faint mimeographed copies of the original log make it very hard to glean any useful information from them.

OVERCOMING A PAUCITY OF INFORMATION ABOUT SMALL VESSELS USING KNOWLEDGE ABOUT TASK FORCE ORGANIZATION

The preceding information may lead to a dilemma for a reader trying to use this book at a vessel level as source about personal family history or a researcher trying to fathom out marine history. If grandpa was aboard one of the two Colberg Boat Works-built rescue tugs at Iwo Jima, and her captain did not keep a war diary, a researcher familiar with the structure of task forces and the reports of actions, might learn at a higher level of reporting about the danger the ship faced, and any engagements with the enemy. I followed this methodology in trying to learn about some ships, when nothing much existed.

For example, Adm. Raymond A. Spruance, commander Fifth Fleet, was in overall command at Iwo Jima. Comprising the Fifth Fleet were a number of subordinate task forces, including Task Force 51, the Joint Expeditionary Force. Within Task Force 51 were a number of task groups, one being Task Group 51.3, the Service and Salvage Group to which the rescue tugs were assigned.

Fifth Fleet Task Organization
Task Force 51 (Joint Expeditionary Force):
Vice Adm. Richmond K. Turner, USN
Task Group 51.3 (Service and Salvage Group):
Capt. Lebbeus H. Curtis V., USNR

If Captain Curtis wanted to further divide his task group, he might assign the rescue tugs USS *ATR-51* and USS *ATR-52* to a subordinate task unit with an associated four-digit designation. If searching for information about *ATR-51*, for example, one would first determine if war diary entries by her were available, or if *ATR-52* compiled such. If the researcher had no luck there, their next step would be to see if other ships assigned to the Service and Salvage Group maintained diaries during the Iwo Jima operation, or submitted reports pertaining to it. From there, they would work upward in the Task Organization and review war diary entries by commander Task Group 51.3 and commander Task Force 51.

By this means, it is possible to piece together the backdrop against which a particular ship operated, in conjunction with or in the absence of any information originated by that ship concerning its operations. This obviously is not ideal, but may be the only option available to researchers and authors. Service members were prohibited from keeping personal diaries during the war, lest they might fall into enemy hands, and letters home were censored.

FATE OF STOCKTON SHIPS, THEIR CREWS AND THE SHIPYARDS

At the beginning of the Korean War in 1950, the strength of the U.S. Navy was seven percent of what it had been at the height of World War II. In the face of the public's desire for a peace dividend, the Navy shed itself of thousands of small vessels that had served in the war. (One exception were some wooden minesweepers that were maintained to continue to clear mines from Japan's inland sea.)

The small ships were gone, the reservist officers and men who served aboard them had been discharged to return to their civilian lives, but sadly most sailors were not sufficiently nostalgic about their former small ships to record their experiences while their memories were fresh. Some stories were undoubtedly told by these men about their service, but less and less often as decades passed. As indicated by the quoted material at the head of this preface, the fine, little ships of Stockton were work vessels, generating little of the public interest lavished on destroyers, PT boats and such. Today, almost none of these little ships remain, and most of the men that took them in harm's way have "crossed the bar." The shipyards and boat yards all along America's coasts that built them are largely forgotten, and their facilities gone—most torn down to allow for more lucrative use of the desirable, picturesque waterfront property which they occupied.

When the war ended, the Moore Equipment Company yard closed and the business was sold to the International Harvester Company. Clyde W. Wood, whose primary business was originally the laying of asphalt and concrete slabs, also ceased building ships. Pollock-Stockton Shipbuilding closed down in February 1946, and the Kyle & Company Stockton facility followed suit in the 1950s. Guntert and Zimmerman bought Hickinbotham's interest in the Hickinbotham yard; thereafter established Guntert & Zimmerman Construction; moved to Ripon, California, in 1984; and continues there in that business. Stephens Brothers, the most significant yacht builder on the West Coast, was in business from 1902 through 1987, producing about 1,200 boats. Colberg Boat Works survived into the 1990s.[18]

With this overview in our wake, it is now time to (vicariously) stand out to sea aboard small ships war bound from Stockton.

1

Patrol Craft Sweeper *PCS-1404* in Action Against Aguijan Island

Upon examination for damage twelve depth charges were observed to have been penetrated by machine gun bullets and/or affected by heat of the burning smoke pot. These were set on safe and were promptly jettisoned in 279 fathoms of water. At 0730 the Commanding Officer of the Marine detachment expressed a desire to return to base to confer with higher authority on the results of the reconnaissance and bombardment; this vessel then took departure from Aguijan Island and proceeded to Tanapag Harbor, Saipan.

—From a report by Lt. William H. Beatty Jr, USNR, commanding officer of the USS *PCS-1404*, regarding his ship's actions against Japanese defense forces on the small island of Aguijan, in the Marianas, on 26 August 1944.[1]

Photo 1-1

USS *PCS-1404* refueling on 5 June 1944 from an unidentifed ship, while en route to Saipan. The tank landing ship USS *LST-277* is in the background. National Archives photograph #193832778

At 0330 on 26 August 1944, the patrol craft sweeper USS *PCS-1404* slipped out of the Charon-Kanoa anchorage at Saipan, Mariana Islands. With a party of Marines embarked on board, she was bound for Aguijan, a small bean-shaped coralline island to the south-southwest, which lay off the the far side of neighboring island Tinian.[2]

Operation FORAGER (February-June 1944), the capture by Allied forces of Japanese-held Saipan, Tinian, Rota, and Guam—all located in the southern end of the Mariana Islands—was officially completed. However, although organized resistance on these islands had ceased, the "mopping up" of remnant enemy groups on them, as well as other smaller islands in the lower Marianas, continued.

Map 1-1

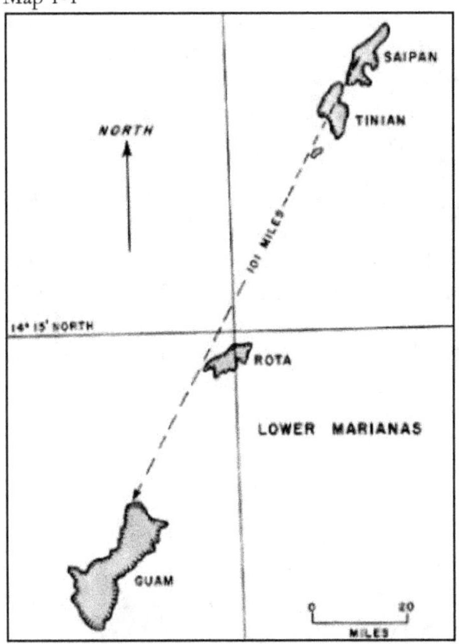

Lower Mariana Islands; Aguijan is the small island to the immediate south-southwest of Tinian

PRECEDING DUTY OF *PCS-1404*

USS *PCS-1404* was a bright and shiny new ship, having been placed in commission less than five months earlier, on 30 March 1944, at Colberg Boat Works, Stockton, California. Over the next 34 days—until sailing from San Diego on 4 May, bound for the Western Pacific—she completed a program of inspection and training to prepare her and her crew for war duty.[3]

Following her commissioning, *PCS-1404* first underwent a ten-day fitting out period, six days at Colberg Boat Works and the remaining four in the San Francisco Bay area. On the morning of 10 April, she stood out to sea, bound for San Pedro, about 450 miles down the California coast. During her transit, she conducted firing tests of her weapons; expending 4 rounds of 3"/50 caliber, 8 rounds of 40mm, and 80 rounds of 20mm ammunition. On 12 April, she arrived at the Small Craft Training Center, Roosevelt Base Anchorage, San Pedro, to start "Shakedown Training," an examination of her material condition, and operational proficiency of her crew.[4]

Map 1-2

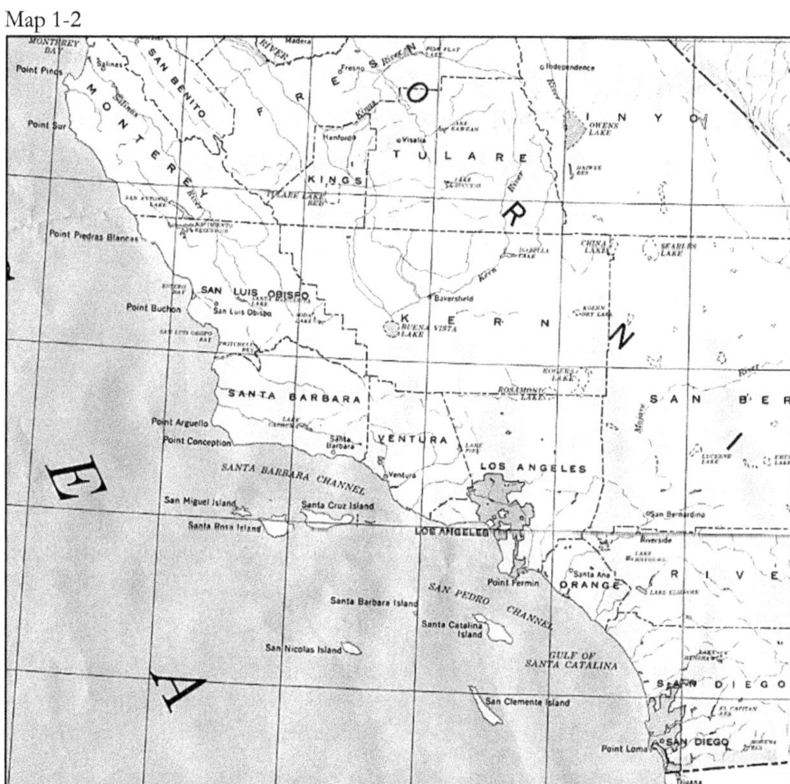

California coastline from Monterey Bay, just south of San Francisco Bay, to San Diego, just north of the United States-Mexico border; San Pedro is a community within the city of Los Angeles, adjacent to the Port of Los Angeles.
https://maps.lib.utexas.edu/maps/united_states/california_south_90.jpg

Over the ensuing 13-day period, *PCS-1404* operated in various training and operating areas in and around San Pedro carrying out a program of instruction prescribed by the commanding officer of the Small Craft Training Center. Shakedown training was completed at 1700 on 25 April. Later that evening the patrol craft sweeper stood out to sea, in company with the steel-hulled minesweepers USS *Clamour* (AM-160) and USS *Scout* (AM-296) en route to San Diego. On arrival there the following morning, she reported to the West Coast Sound School at the San Diego Destroyer Base, for ASW (anti-submarine warfare) training.[5]

Between 23 April and 1 May 1944, *PCS-1404* exercised in various operating areas off San Diego, making mock attack runs on friendly submarines as part of her ASW training. On the nights when she was berthed at the destroyer base, eligible sailors not part of the duty-section required to remain on board for ship safety and readiness, were given Liberty Passes. The one displayed on the following page allowed the bearer to depart their ship, to move freely within the destroyer base, and to leave it to enjoy shore liberty outside the base if they desired.[6]

Photo 1-2

Twenty *Fletcher*-class destroyers at the Navy Repair Base in San Diego, in early 1946 following war's end, while undergoing inactivation (commonly called "mothballing"). Naval History and Heritage Command photograph NH 95684.

Photo 1-3

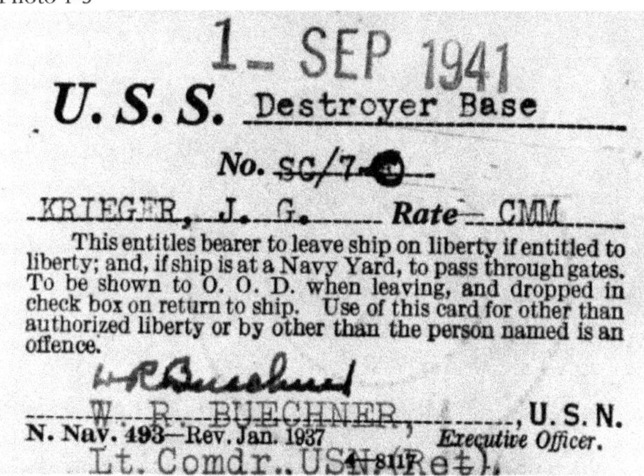

1 - SEP 1941

U. S. S. Destroyer Base

No. SC/7-

KRIEGER, J. G. **Rate** CMM

This entitles bearer to leave ship on liberty if entitled to liberty; and, if ship is at a Navy Yard, to pass through gates. To be shown to O. O. D. when leaving, and dropped in check box on return to ship. Use of this card for other than authorized liberty or by other than the person named is an offence.

W. R. BUECHNER, U. S. N.

N. Nav. 493—Rev. Jan. 1937 *Executive Officer.*

Lt. Comdr., USN (Ret).

Liberty Card issued to CMM John G. Krieger, at San Diego Destroyer Base, California, for 1 September 1941. Signed by Lieutenant Commander W. R. Buechner, USN (Ret). Naval History and Heritage Command photograph #NH 92643

On 1 May her last day of training, *PCS-1404* expended sixteen Mk 22 A/S (anti-submarine) rockets, and seven 300-lb depth charges in structural firing tests. There were no personnel casualties or material damage during this firing exercise. From 2-4 May, she moored at the U.S. Naval Repair Base, San Diego, for a post-Shakedown Availability to fix any new problems discovered during inspection, or continuing ones since leaving the builder's yard.[7]

In early evening on 4 May, *PCS-1404* got under way from San Diego, in company with *PCS-1396* and the minesweeper *Vigilance* (AM-324), bound for Pearl Harbor. When the group arrived there after a uneventful voyage on the morning of 13 May, *PCS-1404* reported to commander, Service Force, U.S. Pacific Fleet. She then received assignment to commander, LST (tank landing ship) Group 8, Flotilla 3, Fifth Amphibious Force for duty. The patrol craft sweeper spent the following week at the naval base provisioning, fueling, and equipping in preparation for forthcoming operations.[8]

At 0636 on 24 May, *PCS-1404* departed Pearl Harbor with four other patrol craft sweepers—*PCS-1396*, *PCS-1457*, *PCS-1460*, and *PCS-1467*—and the yard minesweepers *YMS-151*, *YMS-267*, *YMS-321*, and *YMS-322*, bound for the Hawaiian Sea Frontier Operating Area for short range and anti-aircraft battle practice.[9]

After arriving there, *PCS-1404* made two runs on the short-range target, expending twenty rounds of 3"/50 and thirty-seven rounds of 40mm ammunition.[10]

Photo 1-4

U.S. Navy Gunnery Practice, circa 1913. The target towing and repair ship USS *Lebanon* is towing a target raft with panels erected for short-range gunnery practice by a battleship. Nearby, a tug is turning in the foreground.
Naval History and Heritage Command photograph #NH 101080

In mid-afternoon her training continued. *PCS-1404* next began AA practice with planes towing sleeves on firing runs named George, How, and Charlie-Tare. This exercise involved the ship firing all her anti-aircraft weapons which could bear on the target behind its towing plane, as each run was made. The aircraft supporting such exercises flew a variety of simulated attacks in order to exercise gun crews and fire control parties in anti-aircraft firing and fire control. Of the types of attacks listed below, the practice on that day was against the first three types: glide or dive bombers, strafing planes, and torpedo planes.

- Glide or Dive Bombers (GEORGE)
- Strafing Planes (HOW)
- Torpedo Planes (CHARLIE-TARE)
- High Level Bombing (BAKER)
- Parallel Attacks (UNCLE)[11]

After completion of firing exercises that afternoon *PCS-1404* returned to port at Pearl Harbor and, for the next five days, she continued making preparations for her war mission.[12]

At 0800 on 28 May 1944, she stood out of the harbor, en route to Roi, Marshall Islands, as part of Task Group 51.2 (Northern Defense Group of the Operation against Saipan, Mariana Islands).[13]

The units of this task group are listed below:

LST Group 8, Flotilla 3 Fifth Amphibious Force, Pacific Fleet
Tank landing ship USS *LST-119* (flagship), and
USS *LST-205*, USS *LST-274*, and USS *LST-277*

Escort Group
Destroyer USS *Renshaw* (DD-499)—Patrol and Screen Commander—
and patrol craft sweepers USS *PCS-1396*, USS *PCS-1404*, USS *PCS-1460*

Photo 1-5

USS *LST-119* and other tank landing ships loading the equipment of the US Army Air Corps 804th Engineering Aviation Battalion at Pearl Harbor, 21 May 1944, for transport to Saipan, Mariana Islands.
National Archives photograph 1938345104

Once clear of the port, the tank landing ships formed up, and the escorts took up their screening stations around them. The speed of advance during the transit was a pedestrian 8.8 knots, as the heavily-loaded amphibious ships were capable of only a modest turn of speed. This gave rise to the crews of LSTs referring to their ships as "large, slow targets." The task group arrived at Roi Island, Kwajalein Atoll in the Marshalls, on 8 June, to fuel, provision, and receive fresh water. American forces had captured the atoll in early February 1944.[14]

At 1245 on 10 June, the task group departed Roi, sailing westward for Saipan, Marianas. On the morning of 17 June, the tank landing ships and escort vessels—the destroyer *Renshaw* and three patrol craft sweepers—arrived off Marpi Point, the northernmost tip of the island of Saipan. Rounding the point, they came under fire from enemy artillery emplacements. The fall of the rounds, believed to be 75mm caliber, was short by about 500 yards to the nearest vessel. An evasive course was ordered by the officer in tactical command, and enemy firing ceased after the maneuver was executed by the group.[15]

Map 1-3

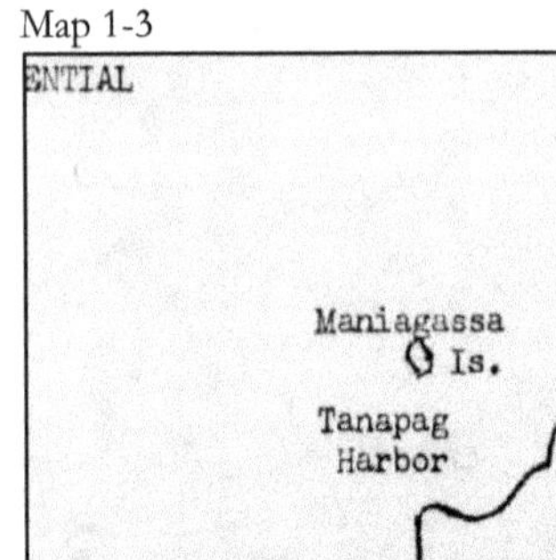

Saipan; map from Commander Group Two, Amphibious Forces, U.S. Pacific Fleet (CTG 52.2) Report of Saipan Operation

At noon, following arrival of the LST formation in the transport area off the town of Charan-Kanoa, commander, Task Group 51.2 detached his screening units. *Renshaw*, *PCS-1396*, *PCS-1404*, and *PCS-1460* then reported to commander, Task Group 52.12 (Transport Screen), and took up an anti-submarine patrol and screen to cover the transport ships at anchor off Saipan.[16]

Details associated with the capture and occupation of Saipan are provided later in the book. After giving the background of how *PCS-1404* came to be in the Marshall Islands, this introductory chapter now returns to the patrol craft sweeper's participation in a little known event, her reconnaissance of Aguijan Island, and resultant exchanges of gunfire with Japanese forces ashore.

EMBARKMENT OF MARINE PARTY

As previously noted in the opening paragraph, *PCS-1404* participated in a special assignment out of Saipan in August 1944. On the evening of

25 August, she embarked Lt. Col. John H. Griebel, USMC, commanding officer of the Aguijan Special Landing Force; Maj. S. D. Mandeville Jr., USMC, operations officer; Capt. John L. Schwabe, USMCR, intelligence officer; Maj. L. E. Haffner, USMCR, commander, RED landing group for the Saipan landings; Lt. W. S. Clarke, USMCR, operations officer, 5th Amphibious Tracker Battalion; and two enlisted Marine radio operators. The group boarded the ship at North Pier, Tanapag Harbor, Saipan, for a single vessel close reconniassance of Aguijan Island, particularly designated GREEN beach on that island, and subsequent bombardment of enemy targets adjacent to the beach. This small island of 2.9 miles in length and a width of 1.1 miles, on which there were Japanese forces, lay only five miles south-southwest of much larger Tinian Island located across the Tinian Channel.[17]

Photo 1-6

Portion of Tinian harbor with Aguijan Island in the background.
Courtesy of Keith D. West via QMCS Larry Richter, USCGR, Ret.

Photo 1-7

Closer, seaward view of Aguijan Island.
Courtesy of Keith D. West via QMCS Larry Richter, USCGR, Ret.

Prior to *PCS-1404*'s departure from Tanapag Harbor, the ship's officers and Marine officers held a conference on board. Using charts, maps, and photographs they discussed the mission—reconnaissance, followed by naval bombardment of identified targets—in detail. In preparation for an invasion following *PCS-1404*'s mission, in preceding days, U.S. Army Air Force P-47 Thunderbolts from Saipan had bombed Aguijan Island. Additionally on 23, 24, and 25 August, units of Escort Divison Sixteen—the destroyer *Cassin* (DD-372), destroyer escort *Cabana* (DE-260), and infantry landing craft *LCI-346* and *LCI-348*—had engaged in neutralizing fire against embedded enemy forces on the island. The shelling of targets ashore was coordinated by voice radio relay from the Second Marine Division on Saipan, which had only one aircraft on scene for spotting and forwarding information. Photographs taken by *Cabana* of the proposed landing beach had been delivered to the headquarters of the Second Marines.[18]

PCS-1404 left Saipan at 0330 on 26 August; arriving two miles to the northwest of Aguijan Island at 0500, thirty minutes prior to sunrise. Despite a first-quarter moon preceding the waning darkness, visability was poor, and her approach to the island was delayed until 0530.[19]

Photo 1-8

Japanese-held Aguijan Island, Marianas, being bombed and strafed in the days preceding this operation in August 1944.
U.S. Army Air Force photograph

Between 0530 and 0625, *PCS-1404* made four slow runs past GREEN beach, located on the southwest coast of the island. The runs were made on course 200°T and reciprocal course 020°T, at an average range of 150 yards from the shore, enabling the Marine officers to examine the beach closely with binoculars and field glasses. During this reconnaissance, the crew was at General Quarters with all guns trained toward the beach.[20]

GUNFIRE BOMBARDMENT OF TARGETS ASHORE

Photo 1-9

Japanese flag flying from structure atop cliffs. This photo, taken after the war, during the surrender of Japanese forces on the island, is likely of the GREEN beach area. Courtesy of Keith D. West via QMCS Larry Richter, USCGR, Ret.

Aguijan Island was a fortress-like rock formation with sheer cliffs all around, topped by the over-grown remnants of a sugar plantation on a plateau. At GREEN beach, there was a small rock or gravel step at water's edge that was used as a landing and paths leading to a small first plateau on the cliffs above, on which were situated four corrugated metal and wood buildings. Helping to make these buildings fast on their precarious perch were two inclined structural steel booms with cables leading down to anchors on the reef near the shore. (These buildings are not shown in the above photo taken after the war. The structure shown appears to be a braced tower from which was suspended a cable way to haul materials and arms from the landing.)[21]

Following the Marine officers' identification of targets, *PCS-1404* opened fire with 3"/50, 40mm, and 20mm weapons, commencing shore bombardment of Japanese installations, machine gun implacements, and geographic points and areas in the immediate vicinity of the beach. Targets and target areas for each of the guns were designated by the Marine officers, with the average range of 1,500 to 2,000 yards.[22]

Orders to the 3"/50 gun crew on the foc's'le came from a spotting officer positioned on the conning bridge, via sound-powered phone communications to the 3"/50 sight setter. The 40mm and 20mm machine guns were on a separate circuit with a talker on the conning bridge conveying orders to them. An officer was stationed aft to serve as a safety observer, to enforce fire discipline, and to "coach" the guns on to their targets. The ship's SO-1 radar, set on the 1-4 mile range, assisted the spotting officer in making more accurate estimates of the proper range to set the 3"/50 gun for a particular target.[23]

At 0700, upon completing a firing run on course 200°T, *PCS-1404* turned around and came to 020°T, for another run along the beach, parallel to it at a speed of 5 knots. On this run, the enemy opened fire from a a triangular cave opening a short distance northwest of the landing step, and from other hidden positions in the green vegetation adjacent to the paths leading up the cliff to the plateau.[24]

In response, the patrol craft sweeper immediately returned fire with all weapons that would bear, including her single 3"/50 gun mount on the foc's'le, and starboard 40mm and two 20mm machine guns. Two 3-inch exploding rounds landed inside the cave opening and burst, creating much smoke. Recurring flashes were observed emitting from the cave as a result of these hits. Meanwhile, enemy machine gun fire began coming from other adjacent areas; bullets were heard passing over the ship, and splashes were observed on her seaward side. In response, the order was given, "engines ahead standard" (12 knots), to increase ship's speed, as all guns continued firing at a rapid rate.[25]

At 0710, *PCS-1404* changed course to open the beach to 2,000 yards in order to get out of the effective range of enemy positions. She then reversed course, and steered 200°T parallel to the beach at 12 knots. All bearable armament opened fire on the enemy positions from which fire was observed to be received on the former run. During this firing run, enemy machine gun fire ignited a smoke pot, stowed as ready service beneath the port depth charge rack. (Smoke pots were used by Navy ships to create a defensive smoke screen to shield them from view of enemy forces, and particularly attacking aircraft.) Fire created by the burning smoke pot was quickly extinguished, and the the ship continued firing automatic weapons until enemy machine gun fire ceased. *PCS-1404* then opened the range to 2,000 yards.[26]

Apart from damage to the smoke pot and depth charges detailed in the quoted material at chapter's head, there was no damge to the ship, or to crewmembers, as a result of enemy gunfire. *PCS-1404* returned to Saipan at 1000 that morning, and for the remainder of August, she was assigned various stations on anti-submarine patrol off Tinian.[27]

AFTERMATH

As previously mentioned, in preparation to capture Aguijan through the landing of troops, a systematic neutralization of island installations by Saipan-based 7th Army Air Force P-47 Thunderbolts had begun on 23 August. For four consecutive days, planes dropped bombs on gun positions, storage areas, buildings, villages, and wooded areas. Following the departure of *PCS-1404* on 26 August, Thunderbolts again bombed and straffed Aguijan. The last mission was on 7 September 1944 when eight planes strafed the island in support of a reconnaissance mission. No enemy activity was observed, and the planned landing of the First Battalion, Second Marine Regiment to occupy Aguijan was abandoned on 9 September.[28]

POST-WAR SURRENDER OF ISLAND

Shortly before noon on 4 September 1945, 2nd Lt. K. Yamada, Imperial Japanese Army, officially surrendered Aguijan to Rear Adm. Marshall R. Greer, USN, commander, Fleet Air Wing Eighteen. Yamada was part of a reinforced infantry company stranded on the island, along with a hundred or so Japanese civilians. The latter were mostly families of the soldiers, who had been part of the garrison on Tinian before the U.S. invasion on 24 July 1944. The ceremony took place on board the U. S. Coast Guard cutter *CG83434* lying off Aguijan Island (see photo on following page).[29]

Photo 1-10

U.S. Coast Guard cutter *CG83434* off Aguijan Island.
Courtesy of Keith D. West via QMCS Larry Richter, USCGR, Ret.

2

Invasion of the Treasury Islands

In summer 1943, the South Pacific Command determined that as part of the final phase of the Solomon Islands Campaign, an assault should be made on Bougainville Island. Occupied by enemy forces, it constituted a major obstacle to Allied forces driving northwest through the New Georgia Group into the northern Solomons. A portion of Bougainville in Allied hands would provide a base for future operations against the powerful Japanese base at Rabaul on New Britain (see map on following page).[1]

As a prelude to combat operations on Bougainville, commander, South Pacific Area, Vice Adm. William F. Halsey Jr., USN, ordered the taking of the Treasury Islands, located to the south of Bougainville. In an operation plan issued 12 October 1943, he directed that Rear Adm. Theodore S. Wilkinson, USN, in addition to other operations, "seize and hold Treasury Islands on Dog minus 5 days [five days before the scheduled landings on Bougainville], capture and destroy enemy forces, and establish threat radars and minimum facilities for small craft as necessary."[2]

Admiral Wilkinson, commander, Third Amphibious Force, split the ships at his disposal for the two landings into the Northern Force for Empress Augusta Bay, Bougainville, and the Southern Force for the Treasury Islands. On 15 October, Wilkinson designated Rear Adm. George H. Fort, USN, as commander, Southern Force. The 8th Brigade Group (less certain detachments) under Brigadier Robert Amos Row were the assault troops for the landings on Treasury. The 8th Brigade was a part of the Third New Zealand Division, which had participated in the campaigns of North Africa, Greece, and Crete.[3]

Third New Zealand Division

Commanding General	Maj. Gen. Harold Eric Barrowclough, DSO
8th Brigade Commander	Brigadier Robert Amos Row
14th Brigade Commander	Brigadier Leslie Potter

Photo 2-1

Adm. William F. Halsey, USN (center) outside the wheel house on board the command ship USS *Appalachian* (AGC-1), in the Marshall Islands, 23 February 1944. With him are Rear Adm. George H. Fort, USN, (left) and Rear Adm. Theodore S. Wilkinson, USN. National Archives photograph #80-G-386436

Map 2-1

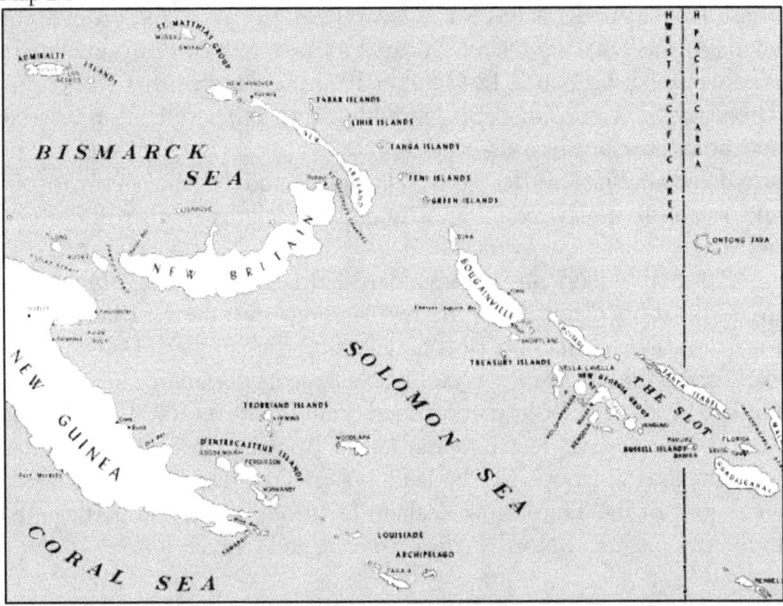

Rabaul Strategic Area (From Rabaul on the northeast tip of New Britain, Japanese ships and aircraft could attack MacArthur's forces working their way up the New Guinea coast, and those of Halsey moving northwest through the Solomon Islands. Rabaul also blocked the planned movement by sea of Allied forces through the Bismarck Archipelago to the Philippines.)
http://ibiblio.org/hyperwar/USMC/II/maps/USMC-II-I.jpg

On the night of 21 October, a reconnaissance party of two New Zealand Army (NZA) non-commissioned officers and some natives were landed by PT boat at Treasury. (The two small islands, Mono and Stirling, which comprised the Treasury Islands were commonly, collectively referred to as "Treasury Island" or "Treasury.") The group learned from local natives that:

- the enemy had recently landed reinforcements
- medium-caliber guns were emplaced on both sides of Falamai Point
- machine guns were emplaced on Mono Island along the approaches to the landing beaches
- there was an observation post at Laifa Point with direct wire communications to a radio station near the Saveke River
- Stirling Island was unoccupied by the enemy[4]

This party was evacuated by PT boat the following night, bringing some Mono Island natives with them to act as guides for the landing. On the night of 25 October, an advance party of NZA non coms and some natives were landed by PT boat on Mono Island. Their mission was to cut the communication lines between the post at Laifa Point and the radio station just prior to the landings on 27 October.[5]

Map 2-2

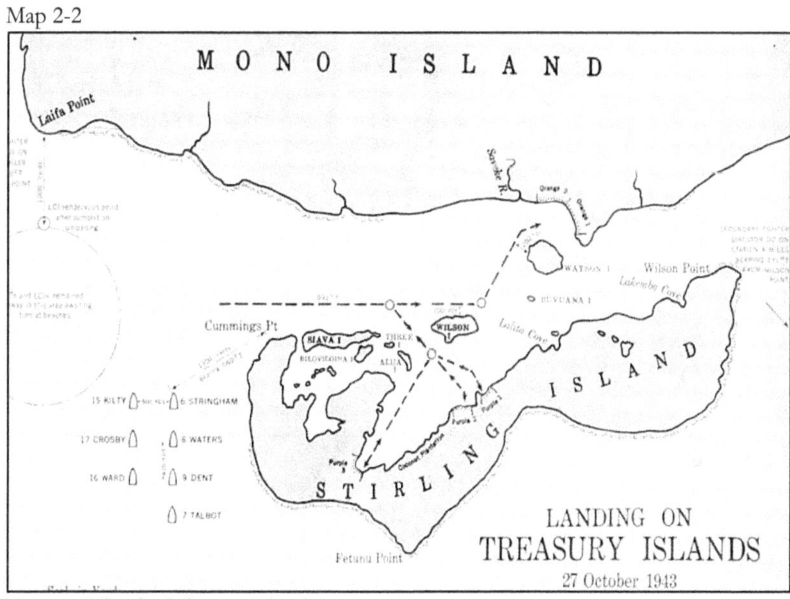

Treasury Islands, Solomon Islands

MOVEMENT TO THE TREASURY ISLANDS

The transport units supporting the attacks and landing on the Treasury Islands under Rear Admiral Fort were divided into five groups, each with their own tactical commander. The groups departed independently for Treasury from the locations indicated with varying times of arrival at Blanche Harbor on 27 October to give each group full use of the beaches, and permit its clearing prior to the arrival of the next group. This was also intended to avoid undue exposure to air attack of vessels awaiting opportunity to unload.

Transports	Departed	Departure Time/Date	Arrival Times on 27 October
1st Group	Guadalcanal	1230 26 Oct	0520
2nd Group	Guadalcanal	0400 26 Oct	0555
3rd Group	Guadalcanal	1930 25 Oct	0640
4th Group	Rendova	1200 26 Oct	0830
5th Group	Lambu Lambu, Vella Lavella	1900 26 Oct	0830[6]

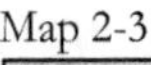
Map 2-3

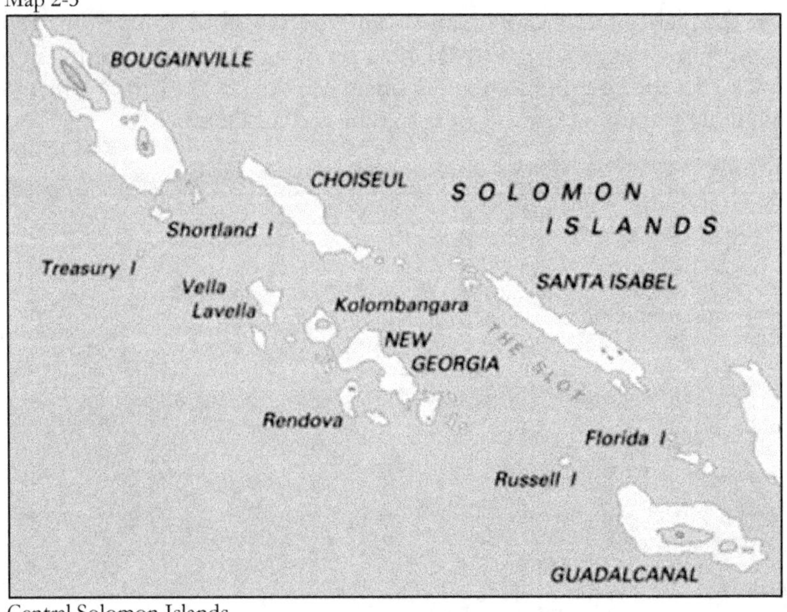

Central Solomon Islands

Eight high-speed transports—*McKean* (APD-5), *Stringham* (APD-6), *Talbot* (APD-7), *Waters* (APD-8), *Dent* (APD9), *Kilty* (APD-15), *Ward* (APD-16), and *Crosby* (APD-17)—and an equal number of infantry landing craft—LCI(L)s *24*, *62*, *67*, *69*, *222*, *330*, *334*, and *336*—made up the first two groups. They were followed by two tank landing ships, *LST-339* and *LST-485*, of the Third Transport Group, and tank landing craft LCTs *321*, *325*, and *330* of the Fourth Group, escorted by the *APc-37*. The final transport group, the *APc-33*, six LCM(3) mechanized landing craft, and one crash boat, entered the harbor last.[7]

STOCKTON SHIP A PART OF THE OPERATION

Photo 2-2

USS *YMS-96* berthed alongside *YMS-355*, location and date unknown.
Courtesy of NavSource

It is noted that the above summary of the composition of transport groups does not identify the escort vessels assigned to each for their protection, with two exceptions. The small 103-foot, unnamed coastal transports *APc-37* and *APc-33* are included because they, like the high-speed transports and landing craft, carried cargo and passengers.

Accompanying the Third Group's tank landing ships *LST-399* and *LST-485* were the destroyers *Conway* (DD-507) and *Renshaw* (DD-499) and the 136-foot wooden-hulled minesweepers *YMS-96* and *YMS-197*. These two small minesweepers were products of shipyards on America's west and east coasts; Stephens Brothers, Stockton, California, and Hiltebrant Dock Co., Kingston, New York, respectively.[8]

Commanded by Lt. Edgar G. Glosson, USNR, *YMS-96*, like Navy small ships generally, was a "maid of all duties." Though her principal role was minesweeping, she also performed escort and anti-submarine patrol duties, armed with one 3"/50 dual-purpose gun, two 20mm anti-aircraft guns, and depth charges. *YMS-96* was obviously not a true sub-chaser. Designed for pulling heavy sweep gear, not speed, her two 880hp General Motors 8-268A diesel engines—coupled to twin shafts via reduction gears—propelled her to a top speed of 13 knots.[9]

The passage of the Third Transport Group (Task Unit 31.3), proceeding northwest from Guadalcanal to the Treasury Islands, was uneventful. A midwatch war diary entry (0000-0400) by *Renshaw* on 27 October gives the formation composition of the ships as the new day begins. The ships were steering base course 320°T, making 10.5 knots. LSTs *485* and *399* were in a single column, the destroyers *Conway* and *Renshaw* were patrolling their stations abreast of the column, and YMSs *96* and *197* were ahead of the formation.[10]

The group arrived off the Treasury Islands at 0700. The YMSs entered Blanche Harbor at 0705, and commenced sweeping ahead of the tank landing ships for all types of mines. The LSTs followed and landed at 0735 on Orange Beach, Mono Island, east of the Saveke River (see map). Five minutes later, both LSTs came under sniper fire and fairly accurate mortar fire, apparently from a position astern. As the destroyer *Eaton* (DD-510) entered the harbor to provide close-in support, mortar fire continued, and hits were made on both LSTs. Small fires resulted but were quickly extinguished. As *Eaton* approached Wilson Island, the firing stopped, with its source remaining unknown.[11]

NEW ZEALAND 8TH BRIGADE ASSAULT FORCES

New Zealand troops went ashore on Mono Island at 0625 on 27 October, in the face of enemy machine-gun fire, and established a beachhead. Enemy resistance against the beach landings was relatively light, and Blanche Bay, the main objective, was taken by nightfall. Nevertheless, by day's end, twenty-one New Zealanders had been killed and seventy wounded. (The landings on Stirling Island, Purple Beach, were unopposed.) Although the Japanese defenders were outnumbered, Mono Island rose steeply from the sea, and dense forest concealed many caves in which the enemy was able to hide. Although complete clearing of the island would prove to be a slow and difficult task; it was nevertheless declared clear of Japanese forces on 7 November, even though isolated enemy positions held out for weeks longer.[12]

Enemy fire directed at the beached LSTs on D-Day was silent for a time, as unloading continued, but then mortar fire resumed at 1125 on 27 October. At 1208, the mortar fire ceased. In mid-afternoon, while screening the tank landing ships, the destroyer *Cony* was hit by two bombs from a Japanese air raid. The blasts knocked out her after engine room and after guns, killed eight men, and seriously wounded ten. The beached LSTs were not attacked, and *Cony* commenced retirement toward the Guadalcanal area on her starboard engine. Other ships were assigned to escort her, and she reached the shelter of Purvis Bay, located south of Florida Island (today Nggela Sule), without further incident.[13]

THIRD TRANSPORT GROUP DEPARTS TREASURY

At 1943 on D-Day, the Third Transport Group completed unloading except for a small quantity of rations aboard *LST-399*, and all units departed Treasury for Guadalcanal to avoid exposure to night bombing while immobile, lacking Allied fighter cover overhead. The two LSTs were escorted by the destroyer *Philip* (DD-498) and the YMSs. At 2024, the formation was illuminated by aircraft flares and float lights (flares that emit smoke) but no bombs were dropped. An hour later, the destroyer *Saufley* (DD-465) joined the formation. At 2257, a float light and two bombs were dropped near *LST-399*, causing minor damage to her radio apparatus. The remainder of the return transit to Guadalcanal was uneventful.[14]

The following day, 28 October, a temporary PT boat base was quickly set up on Stirling Island upon the arrival of Lt. Comdr. Robert B. Kelley and his Motor Torpedo Boat Squadron Nine. Patrols to guard the approaches of the Treasury Islands against counter-attack commenced that evening. With the establishment of this base, PT Boats were also in a strategic position to blockade the areas around southern Bougainville Island, and the Shortland Islands and Choiseul Bay area to the immediate south and east-southeast.[15]

UNIT AWARDS FOR MINESWEEPERS

Each of the five minesweepers that took part in the Treasury Island landing earned a battle star for the day or period identified in the table. The larger steel-hulled, 173-foot AMs *Adroit*, *Conflict*, and *Daring* escorted the 2nd Transport Group to the islands and swept for all types of mines in Blanche Harbor. *YMS-96* was the first of the Stockton, California-built ships to earn a battle star in World War II.

Treasury-Bougainville Operation: Treasury Island Landing

Ships	Award Date(s)	Commanding Officer
Adroit (AM-82)	27 Oct 43	Lt. William D. White, USNR
Conflict (AM-85)	27 Oct-6 Nov 43	Lt. Comdr. Allan D. Curtis, USNR
Daring (AM-87)	27 Oct-6 Nov 43	Lt. William Thomas Hunt, USN
YMS-96	27 Oct 43	Lt. Edgar G. Glosson, USNR
YMS-197	27 Oct 43	Lt. Beckwith A. Brown, USNR

A summary of all battle stars awarded to Stockton-built vessels, associated operations, and periods for which they were earned may be found in Appendix B.

3

Marshall Islands

Map 3-1

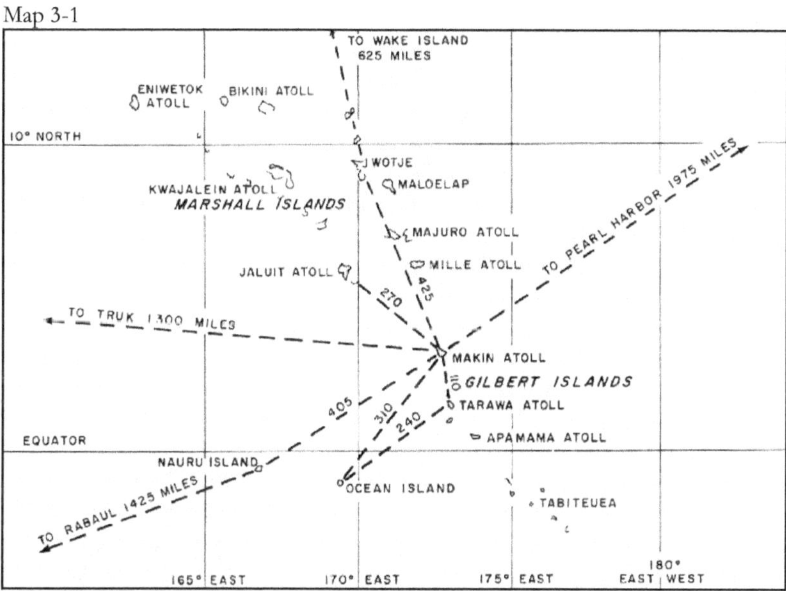

Gilbert and Marshall Islands in the Central Pacific
http://www.ibiblio.org/hyperwar/USN/ACTC/maps/actc-16-p612.jpg

CENTRAL PACIFIC CAMPAIGN OVERVIEW

Allied advances in 1943 included the initiation of a drive by Vice Adm. Raymond Spruance's forces through the Central Pacific on the road to Tokyo, as well as those in progress by Halsey in the South Pacific and MacArthur in the Southwest Pacific. Adm. Chester Nimitz established the Central Pacific Force (Task Force 10) with Spruance in charge on 5 August 1943. Its title would be short-lived. After overseeing the battle of Tarawa in November 1943, Spruance would guide his force as it advanced through the Gilbert Islands, and assaulted Kwajalein and Majuro Atolls in the Marshall Islands. For these successes, Spruance was promoted to admiral in February 1944. Subsequently, the Central Pacific Force became the U.S. Fifth Fleet on 26 April 1944, which Spruance would command through war's end.[1]

Photo 3-1

Adm. Raymond A. Spruance, USN, commander, Central Pacific Force, U.S. Pacific Fleet, on 23 April 1944, days before it was redesignated the 5th Fleet. Navy Photograph, now in the collections of the U.S. National Archives.

The Gilbert Islands, which included Tarawa, Apamama, and Makin, were a group of coral atolls oriented in a roughly north-south line across the equator. The Japanese had invaded the British-held islands on the same day as the attack on Pearl Harbor and occupied them by 10 December 1941. The location of the islands immediately south and east of other important Japanese bases in the Carolines and Marshalls added to their strategic importance to the enemy. For the Allies, the Gilberts offered sites for airfields necessary for progress along the road through the Central Pacific toward Japan. The planned assault and capture of Tarawa and Makin in November 1943 was a part of the overall American strategy of conducting an offensive through Micronesia—the Gilbert, Marshall, and Caroline Islands—at the same time as MacArthur's New Guinea-Mindanao approach to Japan.[2]

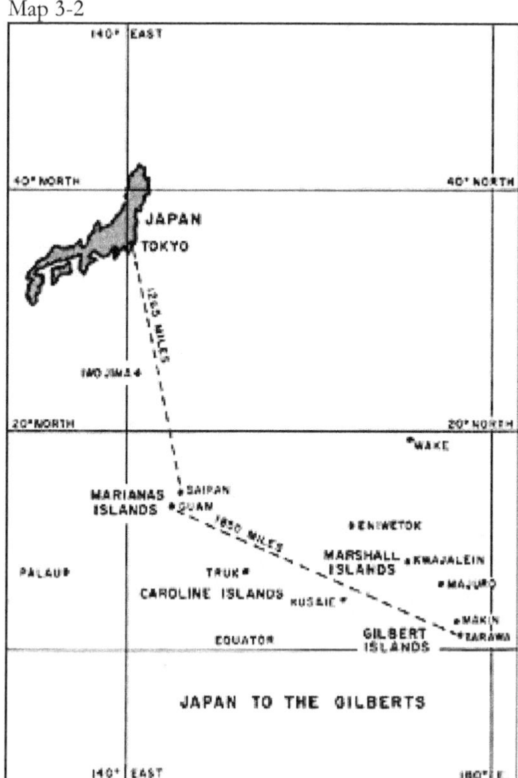

Map 3-2

Principal route of the Central Pacific forces toward Japan
http://www.ibiblio.org/hyperwar/USN/ACTC/maps/actc-p735.jpg

In order to gain airbases capable of supporting operations across the Central Pacific to the Philippines and on to Japan, the U.S. needed to take the heavily defended Mariana Islands. However, use of land-based aircraft to weaken enemy defenses and provide some measure of protection for the invasion forces, necessitated capturing the Marshall Islands, northeast of Guadalcanal. An enemy garrison and air base on Betio, one of the islands of Tarawa Atoll in the Gilberts, guarded against invasion forces arriving from Hawaii. Thus, the starting point for the planned invasion of the Marianas had lain far to the east, at Tarawa.[3]

MARSHALL ISLANDS OCCUPATION

Occupation of the Marshall Islands, which took place over the latter part of January and early February 1944, was a logical step forward in the advance of U.S. forces northward through the Gilbert and Ellice Islands. The probability of the Marshall Islands being the Allies' next objective, must have been known to the Japanese, but uncertainty about the specific objectives was likely fostered by the geography of the area. The Marshall Island group covered an area of roughly 180,000 square miles, consisting of many islands of various sizes. The largest atoll in the Marshalls and in the world is Kwajalein, which has a land area of only six square miles but surrounds a 655-square-mile lagoon.[4]

Map 3-3

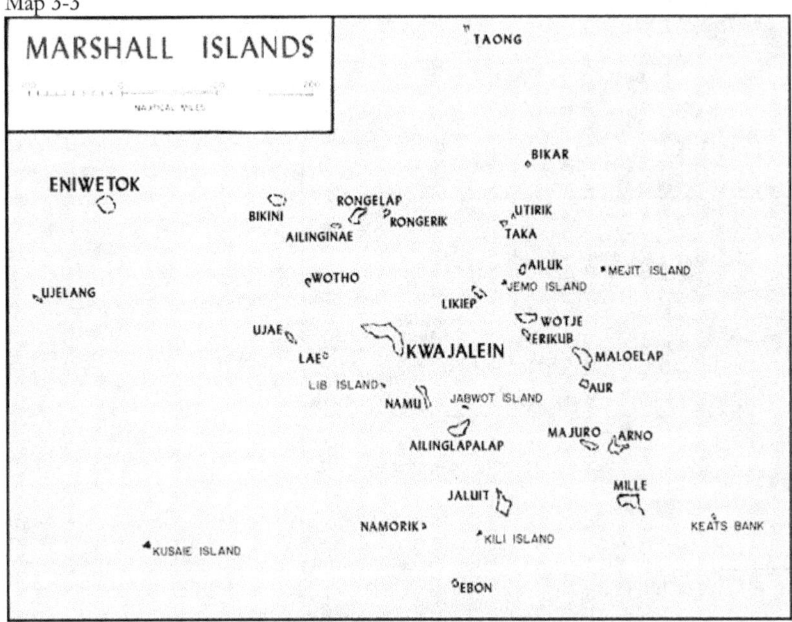

American forces landed on Kwajalein and Majuro during Operation FLINTLOCK, bypassing the other Japanese-held atolls in the Marshall Islands
http://www.ibiblio.org/hyperwar/USMC/III/maps/USMC-III-7.jpg

The general plan was for a Joint Expeditionary Force (Task Force 51) to capture three key points in the Marshalls, in order to establish air and naval bases from which to exercise dominance over the rest of the island group. The objectives were Roi and Kwajalein islands, about 45 miles apart on Kwajalein Atoll—which was believed to be the key Japanese defense point of the Marshalls—and the lightly protected Majuro Atoll, 250 miles to the southeast.[5]

Kwajalein Atoll was a triangular grouping of ninety-three small islands. Because of the vast size of the atoll, Spruance's Expeditionary Force was split into Northern and Southern Landing Forces. The 4th Marine Division was tasked to take Roi-Namur (twin islands considered to be a single island) in the northeast quadrant of the atoll, while the Army's 7th Infantry Division seized the island of Kwajalein itself (44 nautical miles south of Roi-Namur) in the extreme southeast end of the atoll. Majuro Atoll on the eastern edge of the Marshalls, would also be seized in order to provide a fleet base and airfield for subsequent operations.[6]

Map 3-4

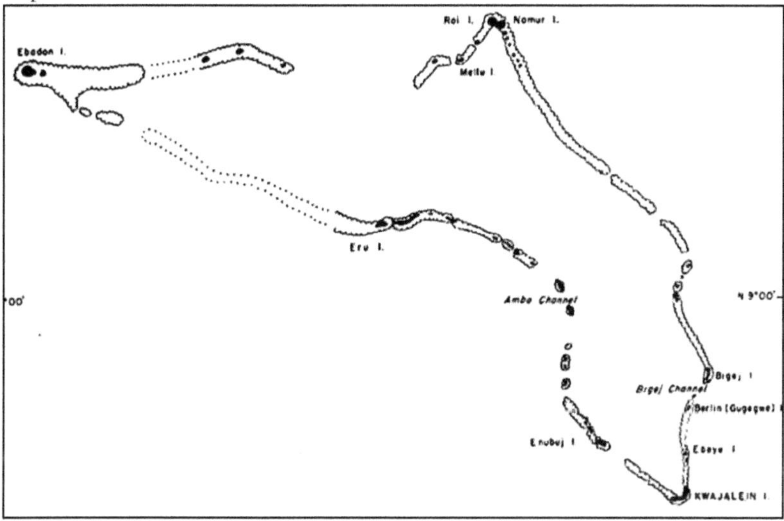

Kwajalein Atoll

This selection decision was Admiral Nimitz's alone. U.S. Navy commanders and planners were cautious after the bloodbath on Tarawa, in November 1943. Vice Adm. Raymond A. Spruance (commander, Fifth Fleet), Rear Adm. Richmond K. Turner (commander, Fifth Amphibious Force), and Marine Maj. Gen. Holland McTyeire "Howling Mad" Smith wanted to take the islands of Wotje and Maloelap in the southern Marshalls before attempting the heavily defended Kwajalein Atoll. After Nimitz overruled them, Spruance and Turner continued to argue the point, until the Pacific Fleet commander, in his gentlemanly manner, offered to replace them if they didn't want to carry out the seizures, dubbed Operation FLINTLOCK.[7]

Vice Adm. Raymond A. Spruance was in overall command, with subordinate task force commanders assigned as shown in the following table. The actual attacks, landings, and seizures of all three objectives were carried out by Rear Adm. Richmond K. Turner's Expeditionary Force. This force comprised some 297 vessels with almost 84,415 troops embarked. Included were transports, troops, and cargo ships, as well as naval support vessels such as battleships, escort carriers, heavy and light cruisers, destroyers, etc.—needed for protection en route, at the objective, and for support of the landings—and a large number of miscellaneous craft.[8]

U.S. Fifth Fleet (TF 50): Vice Adm. Raymond A. Spruance
Joint Expeditionary Force (TF 51): Rear Adm. Richmond K. Turner
Expeditionary Troops (TF 56): Maj. Gen. Holland McTyeire Smith
Carrier Force (TF 58): Rear Adm. Marc A. Mitscher
Neutralization Group (TG 50.15): Rear Adm. Ernest G. Small
Defense and Land Based Air (TF 57): Rear Adm. John H. Hoover
Assault Forces
Northern (Roi-Namur) TF 53: Rear Adm. Richard L. Conolly
Southern (Kwajalein) TF 52: Rear Adm. Richmond K. Turner
Majuro TG 51.2: Rear Adm. Harry W. Hill

Rear Adm. Richmond K. Turner was in overall command of the Expeditionary Force (TF 51), and personally of the Southern Force (TF 52), for the invasion of Kwajalein Atoll. A small component of Task Force 52, were seven minesweepers of subordinate Task Group 52.10.

MINE FORCE

Photo 3-2

Minesweeper USS *Revenge* (AM-110), 6 July 1944.
National Archives photograph BS 78860

The commander of Task Group 52.10, Comdr. Frederick F. Sima, USNR, was similarly "multi-hatted," being also the commanding officer of the group flagship USS *Revenge* (AM-110) and of subordinate Task Unit 52.10.1. This task unit consisted of *Revenge*, a 221-foot *Auk*-class minesweeper, and the two sister ships identified in the table.

Also under Sima's overall command was Task Unit 52.10.2, made up of YMSs *90*, *91*, *383*, and *388*. These much smaller, 136-foot wooden-hulled "sweeps" were under the direct command of Lt. J. H. Pace, USNR, commanding officer of the *YMS-90*.

All seven ships of Task Group 52.10 garnered battle stars for the occupation of Kwajalein and Majuro Atolls, for the period 31 January to 8 February 1944. Two of the members, *YMS-383* and *YMS-388*, products of Colberg Boat Works, were the second and third Stockton-built ships to earn stars in the war.

Marshall Islands Operation:
Occupation of Kwajalein and Majuro Atolls
Task Unit 52.10.1: Comdr. Frederick F. Sima, USNR

Ship	Commanding Officer	Battle Star Period
Revenge (AM-110)	Comdr. Frederick F. Sima, USNR	31 Jan-8 Feb 44
Pursuit (AM-108)	Lt. Romer F. Good, USNR	31 Jan-8 Feb 44
Requisite (AM-109)	Lt. Comdr. H. R. Peirce Jr., USNR	31 Jan-8 Feb 44
Task Unit 52.10.2: Lt. J. H. Pace, USNR		
YMS-90	Lt. J. H. Pace, USNR	31 Jan-8 Feb 44
YMS-91	Lt.(jg) William A. Hurst, USNR	31 Jan-8 Feb 44
YMS-383	Lt. H. C. May, USNR	31 Jan-8 Feb 44
YMS-388	Lt.(jg) R. E. Crowley, USNR	31 Jan-8 Feb 44

TRANSIT FROM PEARL HARBOR TO KWAJALEIN

Departure from the Hawaiian area began with the Assault Force Tractor Groups which, being slower, proceeded ahead. YMSs *90*, *91*, *383*, and *388* left Pearl Harbor for Kwajalein Atoll on 19 January, as escorts for Tractor Unit 1—eight tank landing ships (LSTs), each with three tank landing craft (LCTs) as deck load. Being faster, capable of 18 knots, the minesweepers *Revenge*, *Pursuit*, and *Requisite* departed three days later on the 22nd with the bulk of Task Force 52. During the transit, they were assigned stations in the inner anti-submarine screen. The latter group kept thirty-five miles behind the former en route.[9]

The task force followed a direct line from Pearl Harbor to northwest of the Marshalls. When past Wotje and to the northeast of Kwajalein, it turned to the south-southwest, passed east of Kwajalein, and approached the atoll from the south.[10]

MINESWEEPER DUTIES

Following arrival of the minesweepers at Kwajalein, *Revenge* and the YMSs began sweeping and conducting a hydrographic survey of area "Sweep 1." *YMS-388* was detailed to the survey party until the arrival of landing control craft *LCC-37* and *LCC-39*, at which time she began a magnetic sweep of the area. *Pursuit* and *Requisite* were turned over to the screen commander during the day's operations. That night, the minesweepers took up routine anti-submarine patrols off the passes on the western side of the atoll.[11]

The morning of 1 February, the four YMSs reentered Kwajalein Lagoon via Gea Pass. In preparation for anticipated minesweeping duties, they streamed "O" type sweep gear to port. Recovery of the gear quickly followed, with ensuing time spent provisioning and standing by. That afternoon, *YMS-383* was assigned patrol duty outside the lagoon from Torrutj Pass to Ambo Channel, and the *YMS-91*, to cover northward from Ambo Channel to Burle Island.[12]

At 1600, *YMS-90* joined *YMS-383* in shelling a small Japanese vessel grounded off Eller Island with 3-inch and 20mm gunfire. *YMS-90* then returned to Carloa Transport Anchorage and "dropped the hook" in company with *YMS-388*. Meanwhile, *YMS-91* and *YMS-383* maintained their assigned patrols for the evening.[13]

On 2 February, the four YMSs carried out assigned patrols. The following day, two of the YMSs covered the area from Burle Island south to Torrutj Pass, with another assigned as a pilot vessel on station, and the remaining YMS at anchor in the Transport area in standby.[14]

Pursuit and *YMS-388* in company steamed port "O" gear on the 4th with orders to work the northern half of an area designated as the Southern PORCELAIN Anchorage.[15]

That day, after being relieved on the pilot station by *YMS-388* and en route her patrol station, *YMS-90* sent a boarding party to search a beached Japanese patrol vessel off Gehh Island. It found the vessel to be completely devastated, having evidently been shelled by naval gunfire, with six dead Japanese aboard.[16]

CAPTURE OF ELLER ISLAND

While on routine patrol the following day, 5 February, *YMS-90* received a request by the 158-foot, infantry landing craft gunboat *LCI(G)-438* to assist in the capture of Eller Island with gunfire support. At about 1100, the first troops landed on the south tip of the island from the LCI and an assisting tank landing ship. The troops apparently did not encounter any enemy resistance until advancing halfway across the small island. A sighting was made aboard *YMS-90* of several soldiers emerging from the undergrowth wounded.[17]

Additional tanks were dispatched and within four hours the island was taken. One hundred and one Japanese were killed and one taken prisoner. *YMS-90* shelled a machinegun nest, and thoroughly worked over the remains of a small beached cargo vessel.[18]

SWEEPING COMMENCES

The next three days, 6-8 February, were devoted to routine patrols of assigned areas. On the afternoon of 9 February, *Revenge* (AM-110) received orders to buoy (mark) Ambo Channel. She streamed minesweeping gear to determine if any obstructions were present and, on her first pass through the channel, swept a chemical horn type contact mine. Commander Sima then ordered YMSs *90* and *383* to leave their patrol stations and assist in sweeping the channel. *YMS-383* steamed port "O" gear to a depth of 60 feet, and *YMS-90* magnetic gear. *Revenge* swept two more of the same type mines, and *YMS-383* one. All mines were either exploded or sunk by gunfire, except for one that subsequently beached on Eller Island.[19]

On 10 February, *YMS-90* in company with *YMS-91* and *YMS-388* swept areas of Ambo Channel. One contact mine was swept (cut free from its moor) by *YMS-383* toward the end of the afternoon, and the surfaced mine exploded by gunfire from *YMS-90*. Areas swept clear of mines were marked with Dan buoys.[20]

AIR ATTACK ON ROI ISLAND

The YMSs swept various channels and areas on 11-12 February, encountering no further mines. In early morning on 12 February, a Red Alert (air warning) was announced and aboard *YMS-90*, General Quarters was sounded. An enemy aircraft attack on Roi Island to the north was reported, but no planes sighted, and all hands were released from their battle stations at 0350.[21]

TASK GROUP 52 DISSOLVED

Sweeping operations continued the next few days, during which *YMS-90* swept six moored contact mines on the afternoon of 13 February. All of them were exploded or sunk by gunfire, except the last one swept. It fouled her sweep gear and, found upon recovery of the gear, was cut adrift and buoyed. This mine was sunk by gunfire the following day. Through month's end, no further mines were encountered.[22]

On 14 February, commander, Task Group 52.10 (Comdr. Frederick F. Sima, USNR) learned that effective two days earlier on the 12th, Task Group 52 was dissolved. *YMS-383* was detached for duty with a task group of Vice Adm. Richmond K. Turner's Task Force 51 proceeding to Eniwetok Atoll. On the morning of 18 February, *YMS-90* received notice of the assignment of the remaining three YMSs to commander, Atoll Defense Force, Kwajalein Island. They spent the remainder of the month carrying out routine patrol and pilot duties in the area.[23]

TROOP LANDINGS AND CASUALTIES ASHORE

The landings and combat ashore by the Northern and Southern Forces at the several main and secondary objectives were complicated and difficult. The U.S. troops put ashore—Army and Marines—bore the brunt of the fighting and sustained the major part of the losses. However, the assaults and seizures they carried out could not have been accomplished so expeditiously, and with relatively minor losses, without, as Adm. Chester Nimitz noted in his report on Operations in the Pacific Ocean Areas – February 1944, "a bombardment preparation exceeding in duration and intensity anything previously known to warfare, except possibly that at Verdun in World War I." Nimitz further related in amplification of this statement that:

> It has been estimated that 50% of the Japanese defenders were killed by the air, naval, and artillery bombardments prior to the assaults, not to mention the destruction of defenses and equipment and the stunning effect on the morale and fighting capacity of the survivors.[24]

The expeditionary force that simultaneously invaded both ends of Kwajalein Atoll was massive. It included 54,000 U.S. Marine and U.S. Army assault troops making the initial landings, with gunfire support provided by seven pre–World War II battleships, six escort carriers, and numerous cruisers and destroyers (about 300 ships total). The landings had been delayed from early January to the month's end in order to amass enough assault transports to execute them. Additional air support and cover (and suppression of Japanese bases on other atolls in the Marshall Islands) was provided by six fleet carriers and six light carriers with over 700 aircraft, accompanied by seven modern fast battleships, cruisers and destroyers. Japanese aircraft were swept from the skies before the landings took place and all enemy submarines in the area were sunk.[25]

The 4th Marine Division suffered 737 casualties, including 190 killed, and the 7th Infantry, 177 killed and 1,000 wounded, at Kwajalein. Estimated enemy losses totaled 3,472 dead, with 40 Korean laborers and 51 Japanese the only survivors. Prior to the invasion of Majuro Atoll, Army scouts had canvased Dalap and Uliga islands, found no enemy present, and consequently Majuro was occupied without a fight.[26]

4

Eniwetok Atoll

Eniwetok Atoll (Engebi Island), not highly developed but highly important as a staging base between TRUK and WAKE and the other MARSHALLS fields.

—Bulletin No. 17 Battle Experience: Supporting Operations for the Occupation of the Marshall Islands including the westernmost atoll, Eniwetok, February 1944.[1]

Map 4-1

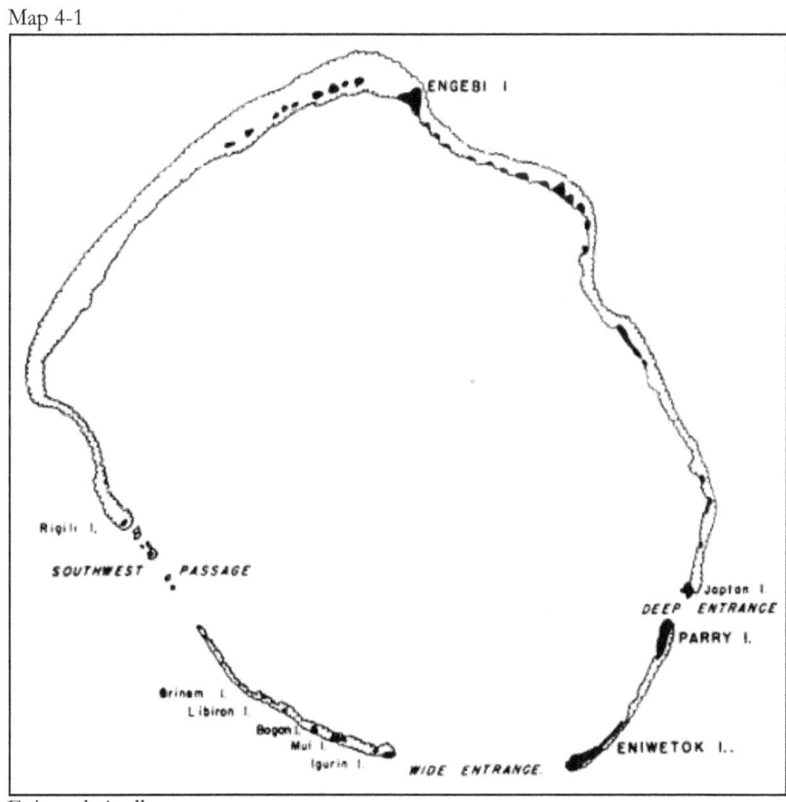

Eniwetok Atoll

Adm. Raymond Spruance's next objective in the Marshall Islands, with Kwajalein and Majuro Atolls taken, was the capture of Eniwetok Atoll. The western outpost of the Marshalls, Eniwetok lay 330 miles west-northwest of Kwajalein. Military commanders intended to use the atoll as an important staging point for U.S. Army, Navy, and Marine Corps forces in their progress from east to west toward Japan. The plan to capture Eniwetok and some lesser Marshall Islands was code named CATCHPOLE.[2]

The three principal islands of Eniwetok Atoll used by the Japanese were Engebi in the northern part, and Eniwetok and Parry in the southern part of the atoll. Engebi Island, the site of the Japanese air strip, was shaped like an equilateral triangle, with each side a little more than a mile long. Eniwetok and Parry were three miles and two miles, respectively in length, and about a mile across at their widest parts. While none of the islands in Eniwetok Atoll attained a height of more than 21 feet, they were the highest in the Marshall Group. This feature allowed the Japanese to construct a defense system of comparatively deep trenches and foxholes.[3]

ENIWETOK EXPEDITIONARY GROUP

The assault forces assembled for the amphibious assault of Eniwetok Atoll was designated the Eniwetok Expeditionary Group (TG 51.11). Rear Adm. Harry W. Hill, USN, commanded the naval force, and Brig. Gen. Thomas E. Watson, USMC, the embarked landing force.[4]

The Minesweeping Group component of the amphibious assault forces consisted of seven ships, assigned to subordinate Northern or Southern Minesweeping Units. All of these ships qualified for a battle star for the period indicated, but were ineligible having previously earned one for the operations against Kwajalein and Majuro Atolls.

Minesweeping Group (TG 51.15)
Northern Minesweeping Unit (TU 51.15.1):
Lt. Comdr. Harry LeRoy Thompson Jr., USN

Ship	Commanding Officer	Period
Chandler (DMS-9)	Lt. Comdr. H. L. Thompson Jr., USN	17 Feb-2 Mar 44
Zane (DMS-14)	Lt. Comdr. W. T. Powell Jr., USN	17-25 Feb 44
Requisite (AM-109)	Lt. Comdr. H. R. Peirce Jr., USNR	17-19 Feb 44

Southern Minesweeping Unit (TU 51.15.2):
Lt. Comdr. John Robert Fels, USNR

Ship	Commanding Officer	Period
Oracle (AM-103)	Lt. Comdr. John R. Fels, USNR	17 Feb-2 Mar 44
Sage (AM-111)	Lt. F. K. Zinn, USNR	17 Feb-2 Mar 44
YMS-262	Lt. Thomas W. Burns, USNR	17 Feb-2 Mar 44
YMS-383	Lt. H. C. May, USNR	17 Feb-2 Mar 44

MINES SWEPT AT ENIWETOK

Diagram 4-2

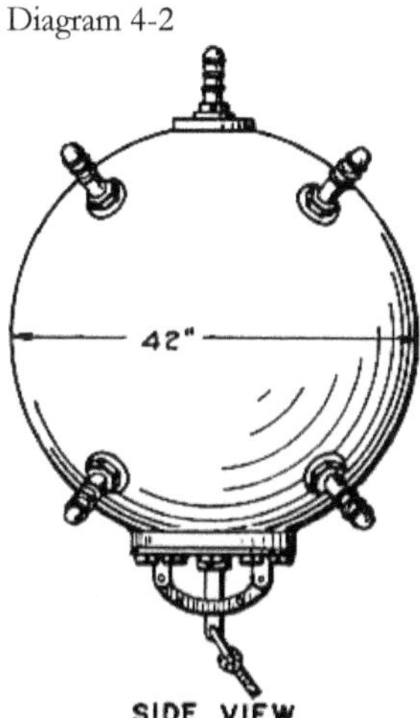

Type-43 Moored, Contact Mine with 9 Lead Horns
Mine Identification Manual, Navy Department Bureau of Ordnance pamphlet, 1943
https://maritime.org/doc/mineid/index.php

Twenty-eight mines were swept at Eniwetok, one each by *Sage* and *Zane*, and twenty-six by *Chandler*. Of these mines, nineteen were detonated on the surface by gunfire, five were observed sinking (as a result of seawater filling their otherwise buoyant casings through bullet hole openings), and four were believed to have sunk.[5]

Zane included a description of the mines, which all appeared to be the same moored, contact-type ordnance, in her report of operations:

> All mines closely resembled type 43 as given in the Mine Identification Manual. Five leaded horns were visible and at times other horns were uncovered when the under side of the mines were exposed in the choppy waters. It is estimated that the mines had nine leaded horns. There was no antenna on any of the mines cut. The mine casing was circular being about four feet in diameter. The majority of the mines were of a reddish hue with little or no fouling. It is estimated that the field was laid within the last three months.[6]

As shown in the diagram from the *Zane*'s report of minesweeping operations at Eniwetok Atoll, one of the mines was found just outside the wide entrance at the south end of the atoll. The others had been laid inside the opening into the lagoon, centered in the entry channel through which shipping was expected to pass.[7]

Diagram 4-3

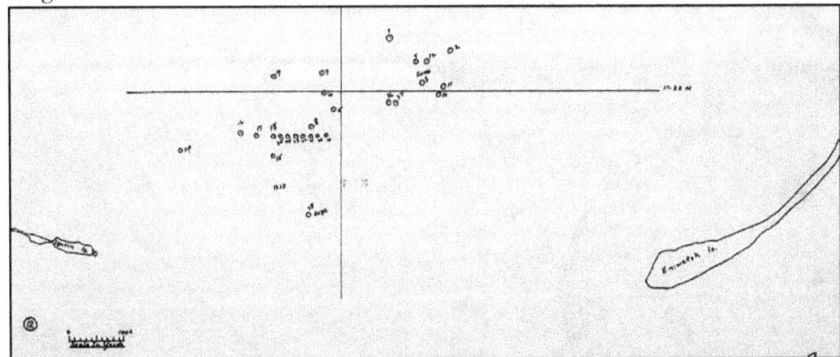

Disposition of mines swept at Eniwetok Atoll

The 314-foot *Zane*, and other World War I-era destroyers converted for the same role, were the Navy's largest and fastest minesweepers. As such, they could clear waters more rapidly than smaller, purpose-built minesweepers, like the 221-foot *Sage* and *Oracle*, and even more diminutive 136-foot, *YMS-383* and *YMS-262*.

Photo 4-1

USS *Zane* (DMS-14) off San Francisco, 21 September 1943.
National Archives photograph #19-N-57504

Photo 4-2

Minesweeper USS *Oracle* (AM-103), location and date unknown.
Naval History and Heritage Command photograph #NH 89204

MAIDS OF ALL DUTIES

While the ships of Task Unit 51.15.1—*Chandler* (DMS-9), *Zane* (DMS-14), and *Requisite* (AM-109)—focused on minesweeping, those of 51.15.2—*Oracle* (AM-103), *Sage* (AM-111), *YMS-262*, and *YMS-383*—carried out more diverse tasking. Many fleet duties that arise, which do not require larger, more valuable ships to perform, are allocated by naval commanders to the so-called "small boys." This occurred at Eniwetok with those of Task Unit 51.15.2, and particularly the YMSs.

Prior to departure of the task unit from Kwajalein on 15 August for Eniwetok, a Hydrographic Unit from the survey ship USS *Bowditch* (AGS-4) embarked aboard *Sage* and *Oracle*. Led by Lt. Comdr. J. C. Tribble, its other members were Lt. (jg) Lucas, and eleven enlisted men together with their hydrographic equipment.[8]

Task Unit 51.15.2 arrived off the atoll with the Southern Eniwetok Expeditionary Group, at 0400 on 17 February, having been used to augment the anti-submarine screen during the transit. Operating independently, the AMs streamed "O" type and "A" type sweep gear, and the YMSs "M" type and "O" type gear, and entered Eniwetok Atoll at 0703 via the Wide Passage. *Oracle* and *Sage* were in a port echelon formation, with sweeps adjusted to twenty-five feet. The YMSs were formed up astern of the AMs in a magnetic sweeping disposition with a station-keeping distance of 400 yards. At 0750, *Sage* swept a moored mine of the chemical contact type, at the southeast corner of a not yet discovered, larger enemy minefield.[9]

GUNFIRE ENGAGEMENT

The Southern Group, led by the destroyer *Phelps* (DD-360), immediately followed the minesweepers into the lagoon, and were joined at a position designated "Point Dog" by the Northern Group which formed up astern. The Northern Group had entered the atoll behind the sweepers *Chandler*, *Zane*, and *Requisite* via the Deep Entrance on the southeast side. The entire force then proceeded to the Transport Area. Upon arrival there, enemy forces aboard a Japanese oiler (AO) beached on Ruunitto Island, were observed firing at a friendly plane.[10]

Upon signal from the commander of the Northern Group, *Sage*, *Oracle*, and *YMS-262* commenced firing their .50-caliber batteries at the oiler at a range of 7,600 yards. Numerous hits were scored, and the enemy gunners abandoned the oiler and took cover on the island. This ship was later shelled by the destroyer *Phelps* and bombed by several aircraft. Discovered aboard the vessel during subsequent investigation were several machine guns (about .303 caliber) and mortars (about 75mm), and two dead Japanese.[11]

YMS-383 DRAFTED FOR SEAPLANE BASE DUTIES

Around noon on D-Day, 17 February, *YMS-383* was released from minesweeping, and proceeded to Aomon Island and took up new duties, as headquarters ship for a seaplane base of limited facilities. The base was established at Eniwetok Atoll in order to relieve parent ships of the trouble of basing their aircraft aboard and reduce fire hazard, while still providing all the seaplane services required. Aomon Island lay in the northeastern part of the atoll, to the southeast of Engebi Island, the site of an existing Japanese airstrip (see Map 4-1). There being no seaplane tender available, *YMS-383* was pressed into this role.[12]

U.S. NAVY SEAPLANE TENDERS

Navy seaplane tenders were employed during World War II to refuel, rearm, and repair the "eyes of the fleet," scout planes and patrol planes, and later in the war, patrol bombers. The ships could be positioned in any sizeable body of protected water where their tended-aircraft could land and take off, and were particularly valuable in advance areas with no facilities for land-based reconnaissance aircraft. Large numbers of the Navy's tenders were devoted to the Pacific Theater during the war, due to the scarcity of existing airstrips or airfields across its vastness.[13]

The Navy's use of seaplanes began in the 1920s, when several World War I-vintage *Lapwing*-class minesweepers were assigned a new mission, caring for seaplanes. These ships, often called "Bird class" because they bore names of birds, received minimum modifications during their conversion to seaplane tenders. Changes were limited to removal of minesweeping gear, addition of storage tanks for aviation fuel, and provisions for berthing and messing aviation personnel. The converted ships retained their AM series hull numbers, denoting minesweeper, for many years until designated "minesweeper for duty with aircraft" on 30 April 1931. The Navy formally reclassified nine ships as small seaplane tenders on 22 January 1936, and assigned them new hull numbers ranging from AVP-1 through AVP-9.

Ex *Lapwing*-class Minesweepers, 187 feet, 1,350 tons

Lapwing AVP-1 (ex AM-1)	*Avocet* AVP-4 (ex AM-19)	*Swan* AVP-7 (ex AM-34)
Heron AVP-2 (ex AM-10)	*Teal* AVP-5 (ex AM-23)	*Gannet* AVP-8 (ex AM-41)
Thrush AVP-3 (ex AM-18)	*Pelican* AVP-6 (ex AM-27)	*Sandpiper* AVP-9 (ex AM-51)[14]

Photo 4-3

Small seaplane tender USS *Avocet* (AVP-4) carrying a Curtiss SOC Seagull scout observation biplane. The former *Lapwing*-class minesweeper still has her original unaltered stack in this photo, which was likely taken in the 1930s.
U.S. Navy photo 1 Hanson Place Brooklyn 17 Navy York
Courtesy of Tommy Trampp

Prior to World War II, the Navy recognized that it would require much greater numbers of seaplane tenders in light of its Pacific Ocean strategy. In the event of hostilities, war plans called for the availability of widely-deployed seaplane detachments to scout for enemy naval forces—and its existing tender force offered only modest capabilities. The small, converted minesweepers could hoist small seaplanes, such as Curtiss SOC Seagull scout observation biplanes and Vought OS2U Kingfisher observation floatplanes, but could not hoist the newer, much larger Consolidated PBY Catalina flying boats. In 1938, the Sea Service began converting the first of fourteen ships drawn from different classes of World War I "flush-deck" destroyers for use as tenders, pending the availability of purpose-built ships. The Navy considered the 314-foot "four pipers" pressed into these duties as too antiquated to serve as front-line combatants. However, being considerably larger than the former minesweepers, they could tend more aircraft, carry greater quantities of aviation fuel, and provide more services for planes and their crews than could the small seaplane tenders. The destroyers were also much faster, and more heavily armed ships.[15]

Photo 4-4

Destroyer seaplane tender USS *Osmond Ingram* (AVD-9) under way off Norfolk Naval Shipyard, Portsmouth, Virginia, on 10 July 1943.
U.S. Naval History and Heritage Command photo # NH 42922

Beginning in 1943, as new, purpose-built *Barnegat*-class seaplane tenders began to replace them, these ex-destroyers began to serve in other roles including convoy escort, anti-submarine warfare, and local patrol, plane guard, and shakedown support for escort carriers.[16]

Photo 4-5

Seaplane tender USS *Barnegat* (AVP-10) under way in Boston Harbor, 1 January 1942, with an OS2U Kingfisher seaplane on her fantail.
Courtesy of Stephen P. Carlson: Boston Navy Yard photo 135-42, Boston National Historical Park Collection NPS Cat. No. BOSTS-10343.

Before leaving this overview of U.S. Navy seaplanes and seaplane tenders, it's important to note that the Japanese also had this capability, and used their seaplanes to scout for Allied forces. The cover art for my book, *Eyes of the Fleet: The U.S. Navy's Seaplane Tenders and Patrol Aircraft in World War II*, depicts two Mavis flying boats carrying out an attack on the seaplane tender USS *Heron* (AVP-2).

Photo 4-6

Evasion of Destruction by Richard DeRosset portrays a strafing run by three Japanese "Mavis" flying boats following their unsuccessful torpedo attack on the USS *Heron* (AVP-2) on 31 December 1942. *Heron* shot down one of the aircraft with her starboard 3-inch gun; her port gun had been disabled by earlier combat action. This final attack followed a series of earlier ones by twelve other enemy aircraft against the seaplane tender as she sailed alone in the Java Sea. Due to heroic actions by her captain and crew, *Heron* survived overwhelming odds during the long ordeal.

ENIWETOK ATOLL SEAPLANE BASE

The seaplane base at Eniwetok was not intended for supporting the larger PBY Catalinas used to scout ahead of the fleet, but instead smaller OS2U Kingfisher floatplanes, or older Curtiss SOC Seagull biplanes, carried aboard the battleships (BBs) and cruisers (CGs) at Eniwetok. Such a base was intended to free the BBs and CGs from the responsibility of recovering and servicing their aircraft every few hours, and thus eliminate interruptions during shore bombardment missions. Although commonly used for both scouting and gunfire spotting missions, these aircraft at Eniwetok were serving only in the latter role.

Air spotters aloft in seaplanes communicated to gunnery officers aboard firing ships whether their gun rounds were "overs" (long), "unders" (short), or left or right of target, and also provided them battle damage assessments of the effectiveness of their naval gunfire.

Photo 4-7

A Vought OS2U-3 Kingfisher being recovered by the cruiser USS *Baltimore* (CA-68) in 1944. These type aircraft were carried by cruisers and battleships for use as scout planes and for spotting naval gunfire.
National Archives photograph #80-G-218123e

Photo 4-8

Curtiss SOC Seagull aircraft on the catapults of a *Brooklyn*-class light cruiser, during operations in the South Pacific area, January 1943.
National Archives photograph #80-G-470115

The seaplane base at Eniwetok Atoll was located in the lagoon at a sheltered point of the east shore about six miles south of Engebi Island. There, *YMS-383* functioned as the seaplane headquarters ship. She hosted the embarked senior aviator of the heavy cruiser *Portland* (CA-33), serving as commander, Seaplanes, and provided berthing and messing for flight crews, despite her diminutive size.[17]

Three boats were utilized for anchorage security patrols and for transportation of flight crews. Eighteen seaplane moorings were laid out, buoyed with empty 20mm ammunition cans because insufficient mooring buoys were available. Planes were fueled and minor repairs performed by the cruisers *Portland* and *Indianapolis*, when they were in the southern part of the lagoon about 15 miles away.[18]

In post-operations reports, ships generally commented favorably about the usefulness of such a base, but strongly recommended that a regular tender be provided, or, if a ship of that type was not available, a dock landing ship (LSD) or tank landing ship (LST) be used. Such a ship acting as a tender, should be able to:

- refuel planes
- make minor repairs
- hoist planes out of the water
- care for plane personnel
- provide sufficient radio communications[19]

Lack of adequate communications aboard the acting tender *YMS-383*, was one of the main deficiencies at Eniwetok. Additionally, she had neither the capability to hoist planes aboard for maintenance, nor perform it.[20]

YMS-262 ORDERED TO CONTROL VESSEL DUTIES

On 19 February, *YMS-262* assumed new duties as a control vessel off Eniwetok Island at the eastern boundary of Wide Entrance, for the next phase of the operation. Assigned to Task Unit 51.14.3, she was part of an amphibious force charged with the capture and occupation of Eniwetok and Parry Islands, which followed that of Engebi Island.[21]

Map 4-4

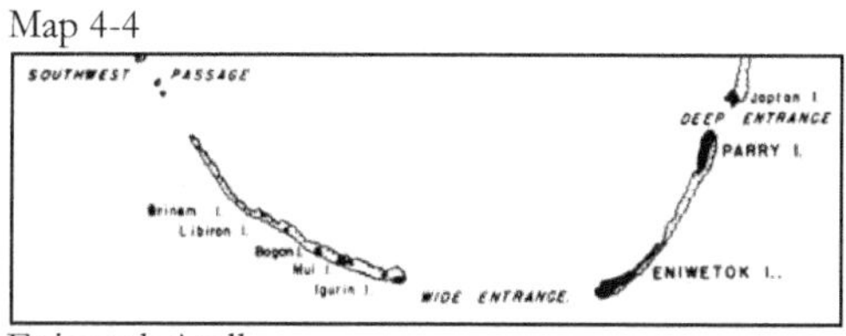

Eniwetok Atoll

Photo 4-9

Coastal ship USS *James M. Gilliss* (AGSC-13), the ex-*YMS-262*, circa 1948-1950. Courtesy of NavSource and Walt Parfenuk

There is little information about *YMS-262*'s activities on D-Day of the Eniwetok Island assault, the 19th, or in the next few days. She secured from her duties as control vessel on the morning of 24 February, and departed Eniwetok Island at 1215 with a group of natives on board. She disembarked them at another island (Clover Island) in late afternoon, and then returned to the Eniwetok Island area and anchored.[22]

The following day, 25 February, *YMS-262* was assigned to an anti-submarine screening station off Wide Pass, which involved maintaining a sound listening watch at all times. The remainder of the month found her and the other minesweepers of Task Unit 51.15.2 assigned to the Gilbert Patrol Group (Task Group 57.7), a surface patrol group of destroyers and destroyer escorts.[23]

CAPTURE OF ENIWETOK ATOLL (CATCHPOLE)

Finally we killed them all. There was not much jubilation. We just sat and stared at the sand, and most of us thought of those who were gone—those whom I shall remember as always young, smiling and graceful, and I shall try to forget how they looked at the end, beyond all recognition.

—Lt. Cord Meyer, USMC, describing the fighting on Parry Island during the assault of Eniwetok. The atoll was taken at a cost of 339 Americans killed and missing, and 2,677 Japanese.[24]

On 17 February 1944, the 22nd Marine Regiment landed at Engebi Island on the north end of Eniwetok Atoll. The thousand or more Japanese defenders offered little resistance following prelanding shore bombardment, and Engebi was taken by late afternoon the next day, at the cost of eighty-five Marines killed.[25]

In southern operations against Eniwetok and Parry islands, the opposing forces were much larger and enemy use on Parry of "hidey-hole" concealment made the combat more difficult. Having learned via captured documents that these islands were defended by 2,155 troops of Maj. Gen. Yoshimi Nishida's 1st Amphibious Brigade, both battalions of the 106th Army Infantry Regiment and a Marine Reserve battalion assaulted Eniwetok Island. Fierce fighting ensued and after much effort, the island was secured at 1630 on 21 February. The cost in American lives was 37 Marines killed and another 94 wounded. The Japanese garrison of about 800 men, except for 23 soldiers taken prisoner, was annihilated.[26]

Photo 4-10

Sailor standing inside a Japanese "rathole" after the capture of Parry Island in 1944. Naval History & Heritage Command photograph #S-487.17

Naval gunfire ships offshore had a lot of time to pound Parry Island before Marines began coming ashore at 0900 on 22 February. Assault troops pushed forward behind tanks, with demolition and flame-thrower parties directly behind to destroy each enemy nest. By that afternoon, the Marines advanced to the island's southern end where the remaining Japanese were in the open. Eniwetok Atoll was taken at a cost of 339 Americans killed and missing, and 2,677 Japanese.[27]

There remained, in the Marshall Islands, many atolls still occupied by Japanese forces, but only four hosted airbases. The Fifth Fleet had bypassed Jaluit, Mili, Maloelap, and Wotje, which could be isolated until war's end through the use of Allied air power. As a part of this effort, patrol aircraft continued to search swaths of ocean, locating and reporting on enemy activity.[28]

5

Patrol Craft Sweepers War Bound from Stockton

The Marianas Campaign, from an amphibious view point had nearly everything; great strategic importance, major tactical moves including successive troop landings on three enemy islands; tough enemy resistance of all kinds, including major Fleet battle; coordination of every known type of combat technique of the land, sea, and air; difficult logistic problems; and the buildup of a great military base area concurrently with the fighting.

— Admiral Richmond K. Turner remarking on Operation FORAGER—
the capture, occupation, and defense of Saipan, Tinian, and Guam—
during a presentation delivered before the U.S. Navy's
General Line School, 5 December 1949.[1]

In early summer 1944, nine patrol craft sweepers arrived in the Western Pacific to take part in a forthcoming Marianas Campaign. The 136-foot, wooden-hulled vessels were all newly-built products of American West Coast boat- and shipyards which were located in the states of California and Washington.

Patrol Craft Sweepers Participating in the Marianas Campaign

California	Washington
Colberg Boat Works, Stockton *PCS-1402, PCS-1403, PCS-1404*	Bellingham Marine Railway and Boatbuilding Co., Bellingham *PCS-1461*
William F. Stone and Son, Oakland *PCS-1421*	Mojean and Ericson Shipbuilding Corp., Tacoma *PCS-1455*
South Coast Co., Newport Beach *PCS-1396*	Tacoma Boat Building Co., Tacoma *PCS-1452*
	Western Boat Building Co., Tacoma *PCS-1460*

Colbert Boat Works, as previously discussed, was at Stockton, located inland from the San Francisco Bay. William F. Stone and Son was in the East Bay region of the San Francisco Bay Area, and South Coast Co., down the coast from the others at Newport Beach in southern California.

The four yards in Washington were at Bellingham and Tacoma, major seaports on the Puget Sound, located inland from the Pacific, and connected to it by the Strait of Juan de Fuca. Tacoma, not shown on the map, lies south of Seattle. Bellingham, well to the north, lies near the United States-Canadian border.

Map 5-1

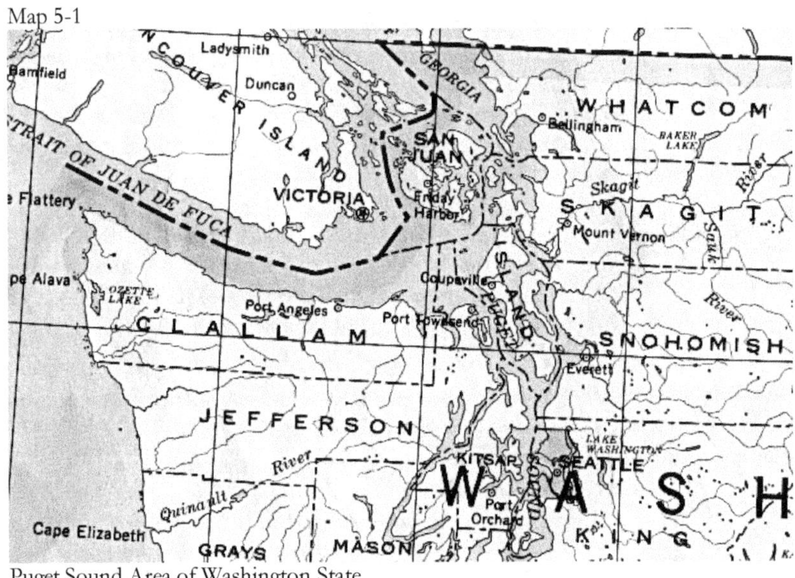

Puget Sound Area of Washington State

SOME PATROL CRAFT SWEEPERS EQUIPPED FOR AMPHIBIOUS CONTROL VESSEL DUTIES

Following their arrival at Pearl Harbor, sailing in groups from the U.S. West Coast, the patrol craft sweepers were assigned to the Fifth Amphibious Force, U.S. Pacific Fleet. For the invasion of Saipan, they had specific duties within Vice Adm. Richmond K. Turner's Northern Attack Force, but carried out others as required. With organizational assignments, as shown in more detail in the following list, their principal responsibilities were:

- Flagship for Capt. Armand J. Robertson, USN, the commander of the Tractor Flotilla (tank landing ships) carrying assault craft for the invasion, and combat materials to land ashore

- Control ships for Commodore Paul S. Theiss, USN (Turner's chief of staff) Control Group directing movements of tank landing ships and assault craft. One of the four PCSs thus assigned served as Theiss' flagship
- Units of Capt. Ruthven E. Libby's Transport Screen, tasked with providing protection to shipping in the transport area

Northern Attack Force (Task Force 52)
Vice Adm. Richmond K. Turner, USN

Tractor Flotilla (Task Group 52.5): Capt. Armand J. Robertson, USN
  USS *PCS-1402* (flotilla flagship)
  Tractor Group Able (52.5.1): Capt. Joseph S. Lillard, USN
    USS *PCS-1403* (Group Able flagship)
Control Group (Task Group 52.6): Commodore Paul S. Theiss, USN
  Central Control Unit (Task Unit 52.6.1): Commodore Paul S. Theiss, USN
    USS *PCS-1452* (flagship)
    USS *PCS-1421*
    Control Group Able (52.6.2): Lt. Raymond J. Koshliek
      USS *PCS-1461*
    Control Group Baker (52.6.3): Lt. (jg) Baxter
      USS *PCS-1455*
Transport Screen (Task Group 52.12): Capt. Ruthven E. Libby, USN
  USS *PCS-1396*, USS *PCS-1404*, USS *PCS-1460*[2]

Turner's large Northern Attack Force (of which the PCSs were only a small part) was assigned the task of invading by sea and occupying Saipan. A separate Southern Attack Force, under the command of Rear Adm. Richard L. Conolly, USN, was to conduct an amphibious assault of Guam, located about 100 nautical miles south-southwest of Saipan.

Map 5-2

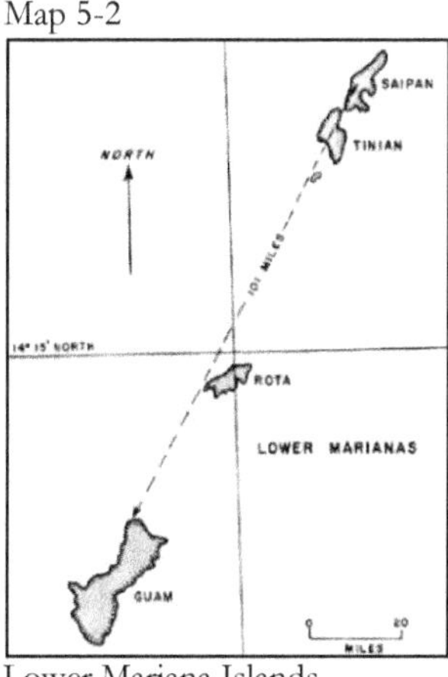

Lower Mariana Islands

To provide more insight into life aboard some of the previously mentioned vessels, the story returns to their departure from the United States. In late afternoon on 17 April, five patrol craft sweepers—PCSs *1403*, *1452*, *1455*, *1421*, and *1461*—had stood out of San Diego Harbor. Once clear of the harbor, they rendezvoused with the attack cargo ship USS *Alcyone* (AKA-7) and formed a protective anti-submarine screen ahead of her. Nine days later, the group arrived at Pearl Harbor, on 26 April. Lt. Willis S. Harrison, USNR, the commanding officer of *PCS-1461*, reported to commander, Fifth Amphibious Force for operational control. He was informed that his vessel was to be used as a control ship in amphibious landing operations, and would receive a large amount of radio gear.[3]

Berthed at Pearl Harbor, his ship subsequently was outfitted and prepared for the special function as control vessel in amphibious operations. On 16 May, she reported to commander, Task Group 52.6 as control vessel for Group Able, with the Control Officer, Lt. Raymond J. Koshliek, embarked aboard. The next several days were devoted to rehearsal of amphibious operations in the area of Maalaea Bay, Maui, and screening tank landing ships in transit to Lahaina Roads, Maui. Afterwards she returned to Pearl Harbor, to have changes made to her radio setup, based on lessons learned in her trials as an amphibious control vessel. On 24 April, she had gotten under way for Maalaea Bay, and the following day proceeded to sea to rejoin the screen surrounding the LST group bound for the invasion of Saipan.[4]

SMALL SHIP CHALLENGES / SHORTFALLS

> *Topped off in fuel from the AO 75 and in water from the YW 76. Came alongside the USS* PRAIRIE *(AD-15) for provisions and small repairs to sound gear. Fuel and water and provision satisfactory in the main, except for fruit. Fifty percent of this issue was spoiled. All escorts had this same trouble. Lay alongside the* PRAIRIE *for the night.*
>
> —USS *PCS-1421* war diary entry related to replenishing from the oiler USS *Saugatuck* (AO-75) and the water barge *YW-76*, and receiving stores from the destroyer tender USS *Prairie* at Eniwetok Atoll. Small ships with junior commanding officers did not always rate the same, as large ships commanded by senior officers.

The assault forces *PCS-1421* was helping to screen stood into the lagoon at Eniwetok Atoll, Marshall Islands, on the morning of 7 June, and all vessels immediately began fueling, provisioning, and making necessary repairs. The next day, the patrol craft sweeper pumped approximately 5,000 gallons of water-polluted fuel overboard outside the atoll on orders of the task group commander. This action was necessitated by a patrol craft sweeper deficiency. They had neither a fuel oil purifier, nor a settling tank to remove impurities including water. The commanding officer of *PCS-1452* nicely described the ongoing problem four months later, in October 1944 war diary entries:

> A great amount of trouble was experienced with injectors and filters in main and auxiliary engines. The injector tips clogged up with dirt and a great amount of water was passing through rather than fuel. The [two] main [propulsion] engines had to be secured at various times to change filter and injectors.
>
> Because of the chance of receiving bad fuel in the forward area, it is recommended that a centrifuge or some kind or a settling tank be installed in these vessels. At present there is no means of ensuring that good fuel is being injected into the engines.[5]

ARRIVAL OFF SAIPAN, MARIANA ISLANDS

Photo 5-1

Naval bombardment of Saipan on 14 June 1944, in support of the planned amphibious assault of the island the following day.
National Archives photograph #80-G-238385

On 9 June, before departing Eniwetok Atoll, *PCS-1461* received a special communications team aboard for the pending amphibious assault operation at Saipan. At 1635, she got under way and joined the screen of Tractor Group Able (LSTs).[6]

Arriving off Saipan in early morning darkness on 15 June, the men aboard the patrol craft sweeper could see fires burning and gun flashes from the island (resulting from prelanding naval shore bombardment) as she began her approach screening the tank landing ships. At dawn, *PCS-1461* sped down the western coast of the island to rendezvous with an APD (high-speed transport), from which were transferred to her naval control personnel for Group Able. Marine Corps tactical and logistical officers and their communications staff boarded shortly afterward as the patrol craft sweeper took station just aft of the line of departure. Often referred to simply as the LOD, this offshore coordinating line was used to assist assault craft to land on designated beaches at scheduled times.[7]

Photo 5-2

Infantry landing craft *LCI(G)-725* and *LCI(G)-726* fire on the beach at Saipan while landing craft form up in the background, during the initial assault, 15 June 1944.
National Archives photograph #80-G-253886

As she went into action, Colberg Boat Works product *PCS-1404* and her crew had only two-and-a-half months service together, since her commissioning on 30 March 1944. It is interesting to note, that sister ship *PCS-1402*, also still relatively green at the time and untested by war, boasted the longest service, her commissioning pennant having been hoisted to her mast top on 13 January 1944.

6

Lower Mariana Islands

Photo 6-1

Marines of the first invasion wave at Saipan hug the beach and prepare to move inland on D-Day, 15 June 1944. An LVT is burning in the background.
National Archives photograph #USMC 81840

The assault and occupation on Saipan launched the larger Mariana and Palau islands campaign, code named Operation FORAGER. On 15 June 1944, six of the nine patrol craft sweepers participating in the invasion of Saipan voyaged from Eniwetok with the tank landing ships carrying assault craft for the initial landings. Upon arrival off the island, they took up duties as flag or control ships for the ship-to-shore movement of the craft landing Marines. The remaining three PCSs arrived on the 17th, and took-up assignments to screen the transport area.

SAILORS ABOARD PATROL CRAFT SWEEPERS

An interesting aside from actual operations is the treatment of PCS personnel concerning wearing of shoulder patches. In June 1944, the Naval Amphibious Forces shoulder patch (shown below at top left) was authorized to be worn by the crews of landing craft that conveyed assault troops to the beach. This badge positioned on the left shoulder above qualified sailors' rating badges, identified them as members of the Naval Amphibious Forces.

The crews of unsung PCSs serving as control vessels at lines of departure during amphibious assaults, and those performing other supportive duties within the range of enemy fire, were not eligible to wear the patch because they did not transport troops.

By the end of 1944, shoulder patches were approved for wear by Minecraft Personnel (top right), Motor Torpedo Boat Personnel (lower left), and pictured at lower right, Construction Battalion Personnel ("Seabees"), but still the PCSs were left out. In January 1947, following the war, the Navy revoked the authorization to wear these patches.

Photo 6-2

Shoulder patches worn by enlisted service members, identifying important, dangerous and, often, little acclaimed functional areas and duties to which they were assigned.

EVE OF ASSAULT LANDINGS

In early evening on 14 June 1944, Capt. Armand J. Robertson, USN, commanding the task group of tank landing ships and escorts, shifted his flag from the destroyer USS *Bailey* (DD-492) to *PCS-1402*, for the approach to Saipan. *Bailey* continued to serve as the fleet center and formation guide for the group. That evening, illumination rounds fired by naval units engaged in shore bombardment were visible, and transport groups were detected by radar to the north.[1]

Continuing throughout the hours of morning darkness, the approach to Saipan proceeded with Tractor Group Baker (Task Unit 52.5.5) in the lead, followed at intervals of five miles by Tractor Group Able, and the Reserve Tractor Group. These units were groups of tank landing ships (LSTs), so designated because they carried LVTs (tracked landing vehicles), a type of amphibious tractor called "amtracs." Commencing at 0530, PCs (patrol craft), SCs (sub-chasers), YMSs (yard minesweepers), and LCI(G)s—infantry landing craft fitted with guns—were detached from the screens of the tractor units to proceed to their assigned duties. The Reserve Tractor Group proceeded to an LST area, seaward of those for attack.[2]

Photo 6-3

An LST launching an LVT amphibious tractor at Okinawa, April 1945, as other LVTs churn by in the foreground. They were taking their positions in the initial assault waves. National Archives photograph #80-G-319224

At 0640, *Bailey* was in position to serve as marker ship for LSTs of Group Baker, and by 0700 all units of that group were in position, opposite beaches Blue 1, Blue 2, Yellow 1, and Yellow 2, approximately 1,500 yards seaward of the line of departure. Also, tank landing ships of Group Able were in position opposite beaches Red 2, Red 3, Green 1 and Green 2. The destroyer USS *Coghlan* (DD-606) acted as a marker ship for that group.[3]

Tank landing ships began launching LVTs immediately upon their arrival in the designated area, those of Group Baker were under the personal observation of Captain Robertson, who circulated throughout the LST area in *PCS-1402*. All LVTs were in the water within eight minutes after the first one was launched with the exception of a single ship, which did not complete her debarkation of craft until 0725.[4]

The pre-landing naval bombardment of the landing beaches was intense. However, despite this, heavy replying enemy mortar fire was observed on the beaches and along a nearby reef as the LVTs transported assault forces shoreward, and continued throughout the day. On several occasions, mortar rounds burst among LVTs, and there was sporadic gunfire directed at the Group Baker LST area.[5]

After the LVTs were launched by the sixteen tank landing ships of both groups, more than half of the ships moved seaward to the rear LST area. Those remaining had been designated to stay at the line of departure with "hot cargo" on call from the troops ashore, or were repair or hospital ships. This cargo consisted of ammunition, food rations, and water. Four ships—LSTs *213*, *223*, *461*, and *487*—were designated to remain as specially equipped LVT repair ships, and three others—LSTs *218*, *341*, and *450*—were fitted out as hospital ships.[6]

Later and continuing throughout the day, other LSTs moved forward to discharge "hot cargo" and the LSTs of the Reserve Group moved to the line of departure to discharge artillery and reserve units.[7]

LANDINGS OPPOSED BY ENEMY FIRE, FOLLOWED BY LENGTHY, HARD FIGHTING INLAND

> *There is something definitely terrifying about the first night on a hostile beach. No matter what superiority you may boast in men and materiel, on that first night you're the underdog, and the enemy is in a position to make you pay through the nose.*
>
> —Statement by 2nd Lt. J. G. Lucas, USMCR, in Navy Department Release of 28 June 1944.

Commodore Theiss and his control officers in the PCSs orchestrated the movements of wave after wave of assault forces, as Marines landed across an expansive beach front, so as to seize a beachhead broad and deep enough to allow deployment inland. Depth was to be obtained by the LVTs (amtracs) giving troops an armored mechanical lift, so that the first assault waves might advance to about a mile behind the beach. This line, in general, followed the 100-foot contour of the foothills, but included 295-foot Mount Fina Susu.[8]

The plan was for subsequent waves to "mop up" behind the first, dig in for the night, and the following day capture Aslito Airfield, advance across the island to Magicienne Bay, and perhaps even capture Mount Tapotchau. The first wave hit the beach at 0845, as scheduled, with the leading elements of the 2nd and 4th Marine Divisions. Within one-half hour, 8,000 Marines were ashore, with another 12,000 on their heels. However, the plan of deep penetration proved unfeasible, because of the skill of opposing Japanese artillery and mortar fire, which forced most of the amtracs to disgorge their troops near the water's edge.[9]

The enemy had numerous extremely-well-concealed artillery and mortar positions on the reverse (back) slopes of ridges and in ravines, where they could not be seen or hit by naval gunfire. Their camouflage had eluded detection by pre-landing aerial reconnaissance efforts, and they would continue to be extremely difficult targets for aircraft to find, let alone hit. The rain of artillery and mortar fire on the beach inflicted heavy casualties on personnel and vehicles. It was soon apparent that Saipan would be no pushover, but instead, a long tough job if it was to be wrested from the Japanese. Advancement forward to the initial planned first-day line, designated O-I, was not fully obtained for three days, and Saipan was not secured for three weeks, delaying the assault on Guam until 21 July.[10]

The Joint Chiefs of Staff had originally set 15 June as the target date for the invasions of Saipan, Tinian, and Guam. Wisely, Vice Admiral Turner—commander TF 51 (Joint Expeditionary Force) and TF 52 (Northern Attack Force)—had left the days for the landings on Guam and Tinian undecided, depending on the turn of events on Saipan.[11]

Photo 6-4

Vice Adm. Richmond Kelly Turner, USN, on board USS *Rocky Mount* (AGC-3) off Saipan, 17 June 1944. National Archives photograph 80-G-231988

VERSATILE SHIPS ASSIGNED WHERE NEEDED

Dropped anchor for the first time in 40 days. Rest was welcome, but it is of interest to note how well a small ship of this type can operate over [an] extended period of time. As more small ships became available, more opportunity to anchor was allowed, otherwise the routine of duties of day and night patrols continued.

—USS *PCS-1396* War Diary entry for 8 July 1944. The patrol craft sweeper was then assigned to the local defense group under the command of Comdr. Ralph S. Moore, USN, who was in charge of the minesweeping group at Saipan.

Patrol craft sweeper versatility and kinship to the yard minesweepers is illustrated by their work together, when not engaged in their respective primary duties. On 30 June 1944, commander, Mine Squadron Four, Comdr. Ralph S. Moore, USN, assumed responsibility for Task Unit 52.12.3 (Inshore Patrol Unit One) at Saipan. This new duty was in addition to his command of the Minesweeping and Hydrographic Group there, which had cleared enemy mines along the east coast of the island and, between 26-29 June, swept Magicienne Bay clear of them. The Japanese, expecting the invasion to take place on the east coast of Saipan, had laid their mines in those areas. The Japanese decision was fortunate. With no mines on the west coast, there had been no impediment to the earlier landings by these type undersea weapons.[12]

Commander Moore was embarked aboard the 221-foot steel-hulled minesweeper USS *Chief* (AM-315), one of the ships of his squadron. The inshore patrol existed for a little over three weeks, being dissolved on 25 July, the day after the invasion of Tinian began. During this period, units assigned to the patrol included:

- Minesweeper USS *Chief* (AM-315)
- Yard minesweepers USS *YMS-291*, *YMS-292*, *YMS-296*
- Patrol craft sweepers USS *PCS-1396*, *PCS-1404*, *PCS-1461*
- Yard net tender USS *Chincona* (YN-7)[13]

In this group, *PCS-1396*'s duties by day were patrolling the transport area off Saipan or the more mundane task of collecting dunnage (ship cargo packaging components) within it. At night, she took up inshore patrols of Japanese-occupied coasts of Saipan and Tinian to detect and destroy any enemy landing barges or torpedo rafts attempting to operate along them. None were sighted and, while the patrol craft sweeper frequently closed the beach of Tinian within enemy range, she did not draw any artillery or small arms fire.[14]

Photo 6-5

Minesweeper USS *Chief* (AM-315) under way in 1952.
Naval History and Heritage Command photograph #NH 96996

On the nights of 1 and 7 July, she sighted and reported low flying Japanese Betty-type planes (Mitsubishi G4M Navy type 1 land-based attack aircraft) which had escaped radar detection. They were observed in the vicinity of the airfield on Tinian, seeking to land. The night of 7 July marked the last of Japanese bombing. On 23 July, *PCS-1396* received orders to get under way as a single escort to convoy the Liberty ship SS *Walter Wyman* to Eniwetok. The passage was uneventful and the pair arrived there on the morning of 28 July.[15]

PCS-1404's experience was similar to that of *PCS-1396*. Between sunset and sunrise from 30 June-23 July, she patrolled one mile off the beaches of various parts of Saipan and Tinian islands to thwart any enemy attempts to escape or gain support from the sea by submarine, seaplane, or small boat. As part of these duties, she maintained an anti-submarine patrol around the transport and seaplane anchorage off Charon-Kanoa town, Saipan, and assisted in keeping the area cleared of stray barges and timbers adrift around the transport area.[16]

PCS-1461 was relieved of her channel control duties on 11 July, and reported to the inshore patrol. She also patrolled close offshore Saipan and Tinian to prevent any enemy troop escapes or landing of supplies by small craft, seaplanes, or submarines. Although she came close inshore to Tinian on her patrols, no shots were fired at her.[17]

SAIPAN TAKEN AT GREAT COST IN LIVES

> *I have always considered Saipan the decisive battle of the Pacific offensive. Saipan was Japan's administrative Pearl Harbor...the naval and military heart and brain of Japanese defense strategy.*
>
> —Lt. Gen. Holland "Howling Mad" Smith, USMC, explaining that the loss of Saipan caused greater dismay in Japan than all her previous defeats put together.[18]

Owing to a stubborn defense of the island by the enemy, Saipan was a costly operation; one resulting in 16,525 U.S. Marine Corps and Army casualties: 3,426 killed in action (KIA), and another 13,099 wounded. Japanese casualties were even higher, 23,811 KIA and 921 prisoners of war taken (including 17 officers).[19]

PATROL CRAFT SWEEPERS EARN BATTLE STARS

Each of the nine PCSs that participated in the capture and occupation of Saipan, earned a battle star, which ships' crewmembers were authorized to affix to the Asiatic-Pacific Campaign ribbon worn on their uniform blouses. This was the first of this type award garnered by USN patrol craft sweepers in the war.

Marianas Operation: Capture and Occupation of Saipan

Ship	Period	Commanding Officer
PCS-1396	17 Jun-24 Jul 44	Lt. Frederic Eugen Sturmer, USNR
PCS-1402	15 Jun-28 Jul 44	Lt. Harry E. Taylor, USNR
PCS-1403	15 Jun-10 Aug 44	Lt. Nelson L. Barnes Jr., USNR
PCS-1404	17 Jun-26 Jul 44	Lt. William H. Beatty Jr., USNR
PCS-1421	15 Jun-10 Aug 44	Lt. (jg) Edmond T. Freeman, USNR
PCS-1452	15 Jun-10 Aug 44	Lt. (jg) W. B. Norwood, USNR
PCS-1455	15 Jun-10 Aug 44	Lt. (jg) Dennis Mann, USNR
PCS-1460	17 Jun-30 Jul 44	Lt. Thomas E. Murphy, USNR
PCS-1461	15 Jun-10 Aug 44	Lt. Willis S. Harrison, USNR

7

Assault and Occupation of Tinian

It certainly was an unusual sight to see those two tiny groups of 8 LVT and 16 LVT [tank landing vehicles] abreast take off from the Line of Departure for beaches White One and White Two respectively. Never had such an insignificant First Wave started a major offensive operation.

—From *The Amphibians Came to Conquer: The Story of Admiral Richmond Kelly Turner*; statement made by Admiral Harry W. Hill to the book's author, Admiral George Carroll Dyer, USN.[1]

Photo 7-1

Marine LVTs head for the Tinian landing beaches on 24 July 1944. National Archives photograph #26-G-2682

Toward the closing of fighting on Saipan, Vice Admiral Turner (commander, Joint Expeditionary Force) directed Rear Adm. Harry W. Hill, USN, to commence preparation of plans for the Tinian Attack Force (TF 52) to capture, occupy, and defend that island. Admiral Hill was then in command of the Western Landing Group of the Northern Attack Force at Saipan. Combat loading operations in support of the planned assault date of 24 July 1944 (designated J-Day), commenced on

15 July in Tanapag Harbor on Saipan, and continued through 23 July. Movement of amphibious ships and craft from Saipan to Tinian, in support of the assault landings and resupply of troops ashore, included:

- 6 APAs (amphibious attack transports)
- 2 APs (transports)
- 2 LSDs (dock landing ships)
- 37 LSTs (tank landing ships)
- 20 LCTs (tank landing craft)
- 31 LCIs (infantry landing craft)
- 92 LCMs (medium landing craft)
- 100 LCVPs (landing craft, vehicle, personnel - "Higgins boats")
- 130 DUKWs (six-wheel amphibious truck – "Duck")
- 537 LVTs (amphibious vehicle, tracked – "Amtrac")
- 14 pontoon barges[2]

Photo 7-2

"Junior Skipper" by G. Richardson, CDR, USN, depicting a Boatswain's Mate 3rd at the helm of an LCVP ("Higgins boat"). Naval History and Heritage Command photograph #NH 118585

Patrol craft sweepers and minesweepers were assigned as part of the Tinian Assault Force. The vessels, their commanders, and the task groups and task units to which they were assigned, are identified below:

Task Force 52 (Tinian Attack Force): Rear Adm. Harry W. Hill, USN
Task Group 56.1 (Northern Troops and Landing Force):
Maj. Gen. Harry Schmidt, USMC
PCS-1421, *PC-1079*, *PC-1080*
Task Group 52.2 (Transport Group): Capt. Clifford G. Richardson
Task Unit 52.2.2 (Control Group): Comdr. Beverly M. Coleman
PCS-1452 (flagship), *PCS-1455*, *PCS-1461*
Task Group 52.5 (Tractor Flotilla): Capt. Armand J. Robertson, USN
Task Unit 52.5.1 (Tractor Group Able): Capt. Joseph S. Lillard
PCS-1403 (flagship)
Task Unit 52.5.5 (Tractor Group Baker):
Capt. Armand J. Robertson, USN
PCS-1402 (flagship)
Task Group 52.13 (Minesweeping and Hydrographic Survey Group):
Comdr. Ralph S. Moore
Task Unit 52.13.1 (Minesweeping Unit One):
Lt. Comdr. Harry LeRoy Thompson Jr., USN
Chandler (DMS-9), *Howard* (DMS-7)
Task Unit 52.11.2 (Minesweeping Unit Two):
Comdr. Ralph S. Moore
Chief (AM-315), *Heed* (AM-100)
YMS-291, *YMS-292*, *YMS-396*
Task Unit 52.13.3 (Minesweeping Unit Three):
Lt. Comdr. John Robert Fels, USNR
Oracle (AM-103), *Motive* (AM-102)
YMS-295, *YMS-385*
Task Unit 52.13.4 (Hydrographic Unit):
Comdr. Ira T. Sanders, C&GS, USN
YMS-296, *YMS-385*, one LCVP[3]

MINESWEEPING AT TINIAN ISLAND

Commencing at 0550 on J-Day, the minesweepers *Chief* and *Oracle* conducted a close-in sweep off White Beaches One and Two, ninety minutes before assault troops were scheduled to land on them. Minesweeper *Motive* followed the others, assigned duties as mine destruction vessel for any moored mines that were cut and rose to the surface. *Chief*, the lead ship, and *Oracle* swept by streaming port gear to 200 fathoms using a 10-foot float pendent, while paralleling the beaches about 350 yards to seaward.[4]

As part of this operation, all three minesweepers raked the beaches and flanking areas with 20mm, 40mm, and 3"/50 gunfire in an endeavor to explode possible land mines—during which ineffective enemy rifle and machine gun fire was encountered.[5]

No mines were swept, and no mines were observed in the area. There were only seventeen sea mines found in the whole Tinian Area, in Asiga Bay, off Yellow Beach on the island's east coast, and these were not cleared until 29-30 July, after the assault landing.[6]

Map 7-1

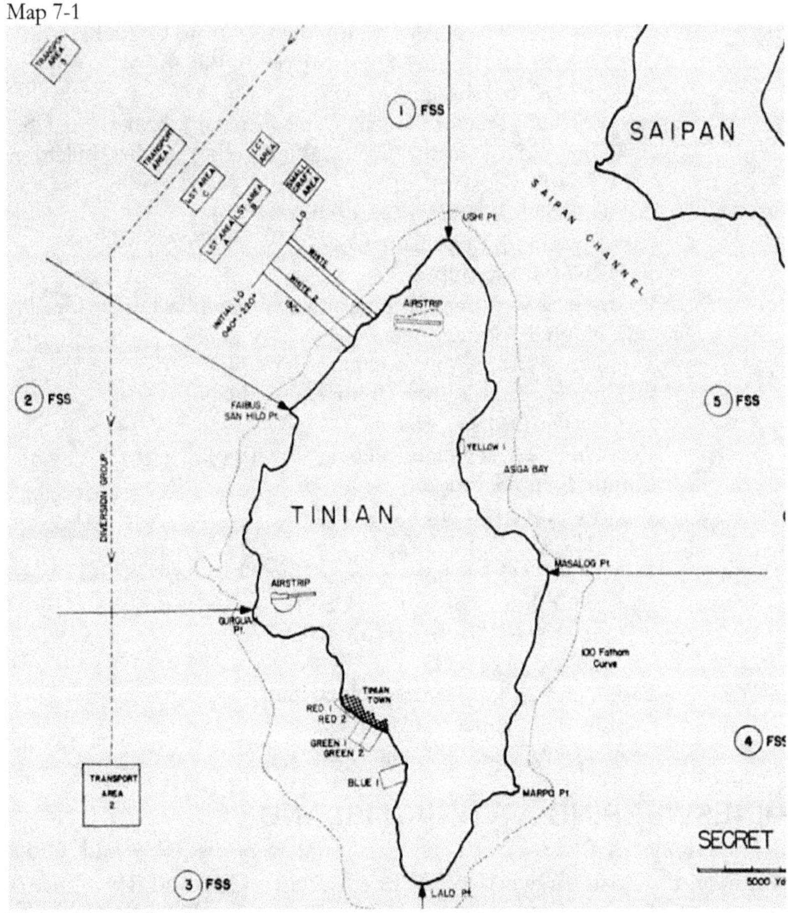

Tinian Assault Beaches
U.S. assault forces landed on White 1 and White 2 beaches on 24 July 1944
Commander in Chief, U.S. Pacific Fleet and Pacific Ocean Area, Operations in Pacific Ocean Areas – July 1944, 22 December 1944

CONTROL SHIP DUTIES

Capt. Clifford Geer Richardson, USN (commander, Transport Division Seven and commander, Transport Group TG 52.2) was responsible for landing troops, supplies and equipment of the Northern Troops on White Beaches One and Two. Tinian had the same type of natural defenses as Saipan's east coast: good sized cliffs and very narrow shallow beaches, which were expected to make the logistic support very difficult in the early hours of the assault landing. On the plus side, in comparison with Saipan and Guam, the area inland of the beaches was fairly flat and fairly open.[7]

Working directly for Richardson, Comdr. Beverly Mosby Coleman, USNR, senior control officer, was tasked with controlling the landing of troops, supplies and equipment of the 4th Marine Division on J-Day (24 July, the day of the assault), and provide the same support for the 2nd Marine Division when ordered. His duties included:

- Synchronizing all assault waves, and advising Richardson of any necessity for delaying H-hour (scheduled time of initial assault landings)
- Embarked aboard *PCS-1452* (the headquarters and control vessel for White One and Two Beaches), maintaining all control vessels on proper station[8]

To execute her duties, *PCS-1452*, commanded by Lt. (jg) W. B. Norwood, USNR, was to receive on board on 23 July, officer representatives of the 2nd and 4th Marine Divisions. On J-Day, at H minus two hours, the patrol craft sweeper was to take, and maintain station on the line of departure (LOD) at a point midway between the boat lanes, identified in the diagram on the next page. When directed by Coleman, flag signals would be used to dispatch the assault waves.[9]

PCS-1461 (Lt. Willis S. Harrison, USNR) was the control vessel for White Beach One. She was to take and maintain station on the left flank of the LOD, at the same time as *PCS-1452* was maintaining overall control, and to follow the movements of the headquarters control vessel in dispatching by flag signals her battalion landing team assault craft.[10]

PCS-1455, under Lt. (jg) Dennis Mann, USNR, the control vessel for White Beach Two, was to take and maintain station on the right flank of the LOD, and follow the same guidance as *PCS-1461*. Both of these ships were to have a Marine Corps representative on board them.[11]

Diagram 7-2

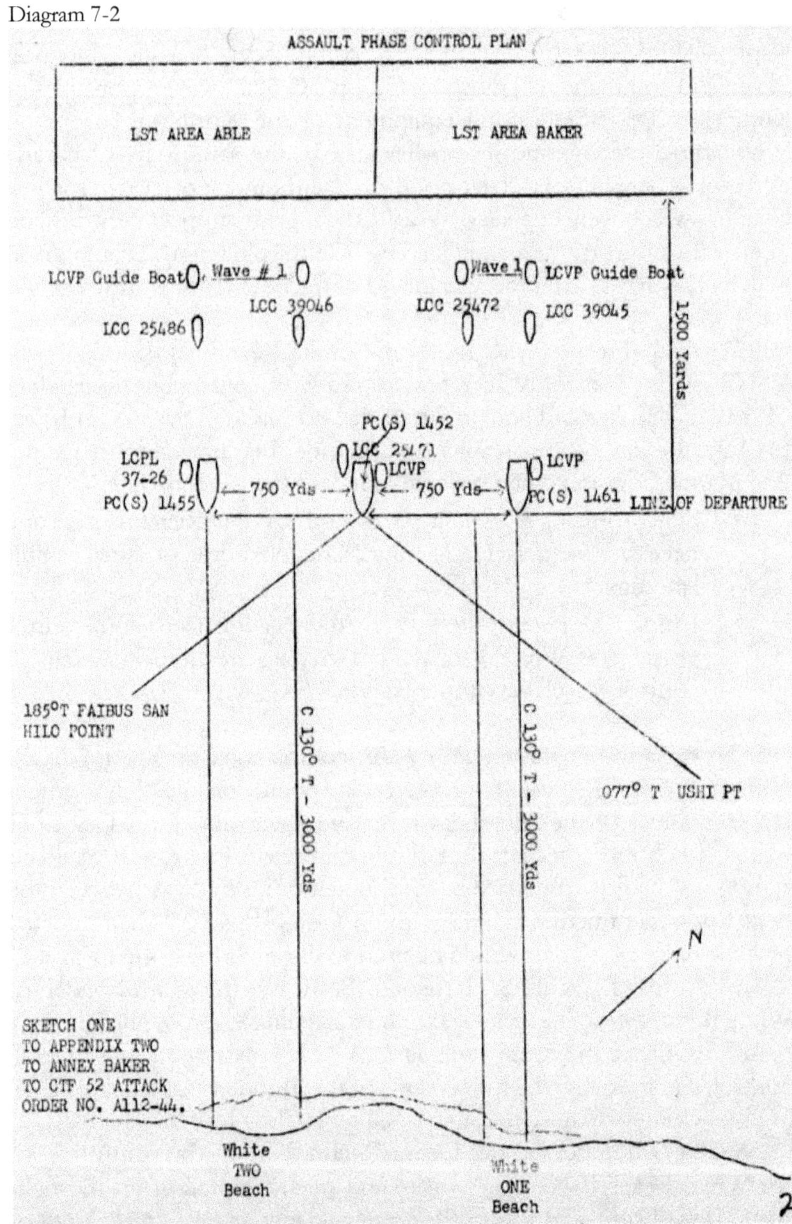

Assault Phase Control Plan; LST Area Charlie is seaward of Able and Baker
Commander Transport Division SEVEN, Report of Amphibious Operations – TINIAN – July 1944, 1 August 1944

Following the establishment of the beachhead, and upon orders of Coleman, *PCS-1461* and *1455* were to move in, and take positions 1,000 to 1,500 yards seaward of their assigned beaches, and serve as primary traffic control vessels.[12]

Diagram 7-3

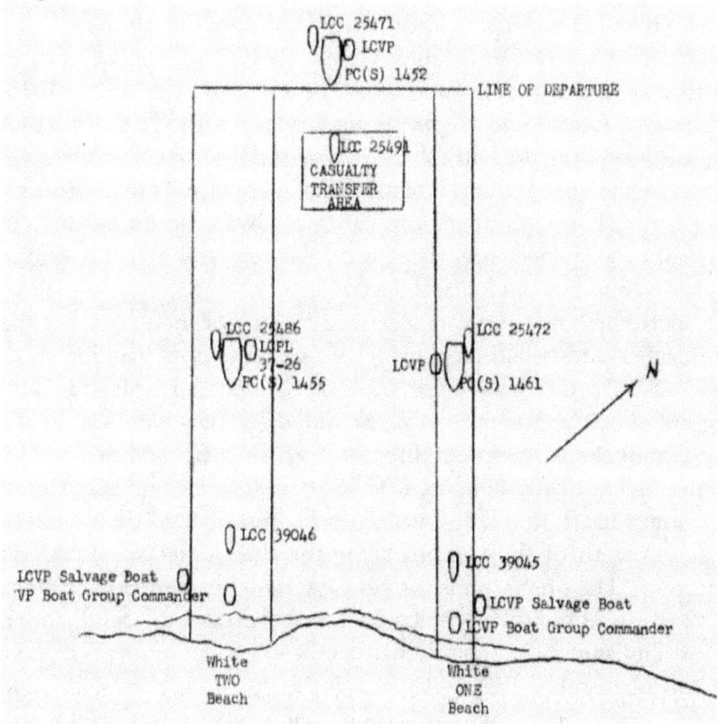

Traffic Control Plan
Commander Transport Division SEVEN, Report of Amphibious Operations – TINIAN – July 1944, 1 August 1944

ASSAULT OF TINIAN

Two additional patrol craft sweepers—*PCS-1403* (Lt. Nelson L. Barnes Jr., USNR) and *PCS-1402* (Lt. Harry E. Taylor, USNR)—served as flagships for the commanders of Tractor Groups Able and Bravo, respectively. (The term "Tractor Groups" referred to amphibious ships carrying tracked vehicles for use in assault landings.) These groups, comprised of tank landing ships, carried LVTs (Amtracs), DUKWs (Ducks), and troops for the assault of Tinian.[13]

Tractor Group Able was under the command of Capt. Joseph S. Lillard, with *PCS-1403* as his flagship; and Tractor Group Baker, Capt.

Armand J. Robertson, USN, with *PCS-1402*. Robertson was also the commander of LST Flotilla Thirteen, from which the tank landing ships were allocated to the Tractor Groups.[14]

On the morning of J-Day (24 July), all movements of shipping to assigned positions in the assault area were executed as planned, and without incident. Naval bombardment, artillery fire, and air strikes began as scheduled in preparation for landings on White Beaches.[15]

Concurrently, a Demonstration Group arrived off Tinian Town to the south, and simulated a landing on the beaches in the area until 1015, at which time it returned to the transport area off White Beaches. During the deception, the battleship *Colorado* and destroyer *Norman Scott*, which were executing fire support missions, were heavily hit by rounds from a previously undetected shore battery. Details about this action, and casualties sustained, are discussed following the account of the landings at White Beach One and Two.[16]

At 0600, the Tractor Groups arrived in the assigned areas off the White Beaches and began launching LVTs, as the dock landing ships *Ashland* (LSD-1) and *Belle Grove* (LSD-2) launched LCMs (medium landing craft) carrying tanks. *Ashland* and *Belle Grove* were essentially floating dry docks carrying amphibious craft in their well decks. To launch them, they "ballasted down" (filled ballast tanks with seawater to increase ship's draft, flooding the well deck), and opened the well deck doors at the stern of the ship, to allow the then waterborne craft to proceed out. The attack transports—*Cambria* with commander, Task Force 52 (Rear Admiral Hill) embarked, and *Cavalier* with commander, Transport Division 7 (Captain Richardson)—took station near the LST area. Control vessels—*PCS-1402*, *PCS-1403*, *PCS-1452*, *PCS-1455*, and *PCS-1461*—took their assigned stations as previously described.[17]

Emplaced on the south coast of Saipan were ninety-six 105mm howitzers, thirty-six 155mm howitzers, and twenty-four 155mm guns, supporting the assault waves forming offshore, by directing fire in the vicinity of the landing beaches. Two battleships, one heavy cruiser, and four destroyers also delivered fire in support of the landings, lying to in assigned stations close to the beaches and the boat lanes. At 0628, gunfire ceased allowing pre-landing aircraft bombing and strafing attacks on the beach area.[18]

The first assault wave scheduled to land assault troops was dispatched at 0717. Thirty LCIs (infantry landing craft not involved with transportation of the troops) supported the landings with rockets and 40mm gunfire. Part of these led the LVT assault waves to the beaches. The remaining LCIs moved in toward the beaches, outside the boat lanes, and delivered close support fire on the flanks. In this operation, *LCI-460* was hit by enemy gunfire, seriously wounding two members of her boat crew.[19]

It was known that the usable width of White One Beach was only about 60 yards, and of White Two about 120 yards—the remainder of the beach areas were obstructed by coral boulders and ledges. It was also known that there were anti-tank mines emplaced on White Two Beach. Accordingly, troops were instructed to disembark from LVTs at the water's edge, and make their way inland between boulders. All other LVTs were instructed not to land until beaches were cleared of mines. Engineers successfully removed fifteen anti-boat mines from this beach, all buried between high and low water marks. In spite of these precautions, two LVTs were blown up on White Two Beach.[20]

The assault landings were opposed by sporadic rifle fire and some machine gun fire from inland locations, but resultant losses were extremely light. Most of the fifteen deaths on the first day of the assault were associated with the two destroyed LVTs. The commanding general, Fourth Marine Division (Maj. Gen. Clifton B. Cates) reported satisfactory progress, with 48 medium tanks and four battalions of 75mm artillery ashore by day's end. Although opposition was light, numerous land mines and booby traps were encountered. Casualties were 15 killed, and another 255 wounded.[21]

AMPHIBIOUS DEMONSTRATION OFF TINIAN TOWN

> *Admiral Turner wanted to land on the good beaches (strongly defended) at Tinian Town. The Landing Force wished to land on the narrow and very poor beaches (relatively undefended) near the north end of Tinian. The latter beaches could be dominated by our ready placed artillery on the south coast of Saipan, and we were convinced we could negotiate the beaches as a result of our UDT [underwater demolition team] and beach reconnaissance….*
>
> *On the last possible day of decision General Holland Smith, his CIS [Communications and Information Services Officer], his G-2 [Intelligence Officer], and I went out from Saipan to Admiral Turner's flagship expecting a knock down fight. We were prepared to press our plan in great detail and in the strongest terms. As soon as our plan was stated in outline, Admiral Turner turned to General Smith and said: 'I can support your plan, I approve.'*
>
> —Recollections of the operations officer for the Landing Force commanded by Lt. Gen. Holland McTyeire Smith, USMC, during the Saipan-Tinian Operation, regarding the ultimate selection of White One and White Two Beaches for the assault on Tinian Island.[22]

Following much debate, at times heated, among top U.S. military officers regarding the selection of landing beaches at Tinian, the White Beaches were chosen. There were only three possible landing areas on Tinian, as shown on the preceding Map 6-1: Sunharon Bay on the southwest coast of the island, Asiga Bay on the east central coast, and the White Beach area in northwest Tinian.[23]

The two big advantages of the northwest "White" beaches were that they were within range of direct artillery support from Saipan; they were on the sheltered, leeward side to prevailing winds of Tinian; and near an airfield whose rapid capture was necessary. The tremendous disadvantages were their extremely small frontage, and their narrow steep exits. Division commanders landing troops generally desired a beach with a mile or more of frontal access, with many exits, to preclude assault forces having to advance inland in the face of enemy defensive fire, while "bottled up" with few egress options available to them.[24]

Admiral Turner's ultimate decision was insightful. The Japanese were fully aware that the beaches off Tinian Town, in Sunharon Bay, south of the White Beaches, were inviting. Aerial reconnaissance as J-Day approached revealed they were making real last-minute efforts to improve their defensive stance in this, and the Yellow Beach area in Asiga Bay on the island's east central coast. Their anticipation was further confirmed later, following the landings, by a captured order of the Tinian Garrison force dated 1900 on 25 June, which indicated that the U.S. attack was expected at the Tinian Town and Asiga Bay areas.[25]

On J-Day, concurrent with the landings at the White Beaches, Task Group 52.8 conducted a diversionary action off Blue Beach, south of Tinian Town, in order to deceive the Japanese regarding the primary point of attack and to immobilize enemy reserves sent to defend against the amphibious forces. Commanding the task group was Capt. Clinton A. Misson, USN. The deception was staged but quickly withdrawn. Upon completion of the feint, the ships were diverted to Transport Area Three, seaward of the White Beaches and their resources used to support the main assault. Landing Team 1/8 was disembarked from the attack transport *Calvert* (APA-32) on J-Day, and the remaining troops and equipment on J+1 Day; all on White One Beach.[26]

BATTLESHIP USS *COLORADO* AND DESTROYER USS *NORMAN SCOTT* HIT BY ENEMY FIRE, SUFFER MANY CREWMEMBERS KILLED AND INJURED

During the entire operation we were impressed with the ability of the Japanese to conceal their installations and personnel. In spite of the large numbers of people on SAIPAN, TINIAN, and GUAM, few were seen from this ship. On the other hand, our troops, tanks, and motor transport, could frequently be seen.

—Observation made by the commanding officer of the light cruiser USS *Cleveland* in an action report, under the heading, "Ability of the Japanese at Concealment."[27]

Enemy shore batteries and artillery were suspected of being emplaced in the vicinity of Tinian Town, together with other defensive measures to counter any attempted landings there. Consequently, at 0610 on 24 July, when commander, Task Group 52.8 (Demonstration Group), ordered "Land the Landing Force," and all boats engaged in the demonstration were lowered into the water, fire support was being carried out by a pre-H-hour air strike and naval gunfire to lend credibility to the operation. (H-Hour is the designated time at which the first landing craft or amphibious vehicle of the waterborne wave lands or is scheduled to land on the beach.) The latter was delivered by the battleship *Colorado* (BB-45), light cruiser *Cleveland* (CL-55), and the destroyers *Remey* (DD-688), *Wadleigh* (DD-689), *Norman Scott* (DD-690), and *Monssen* (DD-798).[28]

The Japanese had heavily fortified Tinian with coastal defense, anti-aircraft, and machine guns. Practically all the coastal defense guns, many anti-aircraft guns, and a large portion of other defenses had been destroyed before the amphibious landings. However, a skillfully camouflaged battery of three 6-inch guns, located at the base of caves near Tinian Town, had escaped detection and destruction. There also remained mobile artillery and mortar pieces.[29]

The Demonstration Group, consisting of five attack transports (APA) and two transports (AP) with two regiments of the 2nd Marine Division embarked, together with control vessels *PC-581*, *PC-582* and other vessels, had arrived off Tinian Town beaches at around 0600. Boats were lowered and embarkation of troops simulated, with Marines descending debarkation nets, but not entering the boats. The boat group, consisting of 100 empty LCVPs ("Higgins boats"), formed assembly circles, then approached the line of departure.[30]

The first wave, coordinated with the actual assault on White Beaches, left the LOD at about 0717. When this wave was about 2,000 yards from the beach, intense enemy mortar and artillery fire broke out. This wave, and successive ones, continued shoreward for about 1,000 yards more; then, on a pre-arranged signal, countermarched (turned around) and returned to seaward. There were no casualties to personnel or to the boats taking part in the demonstration.[31]

Photo 7-3

USS *Colorado* (BB-45) in Puget Sound, Washington, on 25 April 1944. Bureau of Ships photograph #1426-44

At about the same time that mortar and artillery fire opened, the battleship *Colorado* and destroyer *Norman Scott* were taken under fire. The following excerpt from an action report by the destroyer *Remey* describes a duel between the enemy battery and her naval gunfire:

> 0740 boat waves passed between this ship and U.S.S. *NORMAN SCOTT*.
> 0741 U.S.S. *COLORADO*, on port quarter received several hits on starboard side amidships and starboard side seemed to be aflame, U.S.S. *NORMAN SCOTT* observed to be under fire. Commenced maneuvering.
> 0742 battery of three guns located dead astern in wooded area…observed firing on *COLORADO*, these guns were firing very rapidly and appeared at first to consist of more guns than that determined at a later date.
> 0743 commenced firing with after guns.
> 0744 all guns firing.
> 0746 checked fire momentarily while passing to seaward of *COLORADO*, who was still under fire, and resumed fire immediately thereafter.

> 0750 enemy coastal battery covered by dense smoke and only one gun salvo observed still firing at *COLORADO*... Reduced rate of fire.
> 0807 ceased firing having expended 274 rounds AAC projectiles and service powder in counterbattery fire.
> 0825 commenced screening U.S.S. *COLORADO* which had received many direct hits. The marksmanship and firing of this enemy battery was superb.[32]

The *Colorado* was hit 22 times by a battery of six-inch coastal guns, resulting in 178 casualties including two officers and 42 enlisted men killed. The *Norman Scott*, also badly hit, suffered 19 killed, including her commanding officer, and 47 wounded.[33]

CAPTURE OF TINIAN

Diagram 7-4

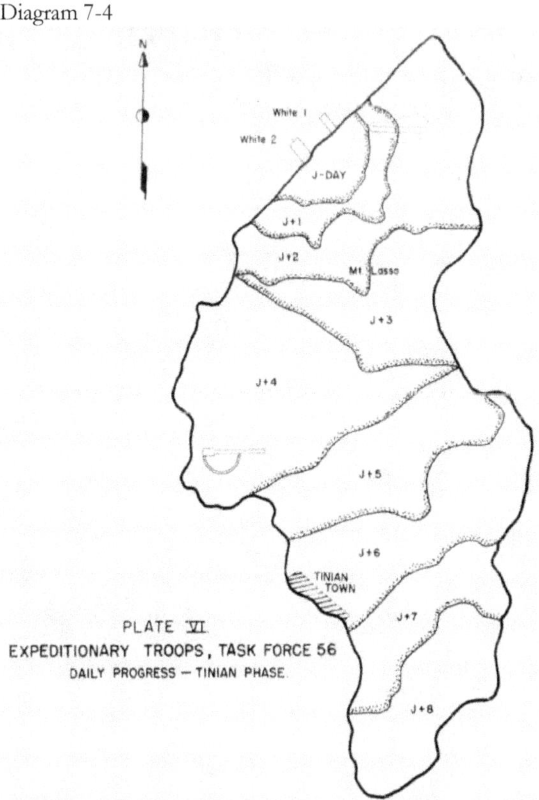

Daily progress by ground troops ashore at Tinian
Commander in Chief, U.S. Pacific Fleet and Pacific Ocean Areas, Operations in Pacific Ocean Areas – July 1944, 22 December 1944

The assault over White Beaches apparently caught the Japanese island defense force by surprise. The U.S. Marines moved fast south-southeast down the island, as shown by the above diagram, allowing the enemy no opportunity of repositioning to meet an attack from the north. On the evening of 1 August, the island was reported secure.

The final numbers of American and Japanese casualties reported to commander, Task Force 52, were as follows:

- American KIA: 290, WIA: 1,515, MIA: 24, total: 1,829
- Japanese buried 5,524, POW: 404, civilian casualties: 13,262

PATROL CRAFT SWEEPERS EARN BATTLE STARS

All of the patrol craft sweepers at Tinian earned battle stars, except for *PCS-1421*, which General Schmidt employed for administrative duties.

Ship	Award Period	Commanding Officer
USS *PCS-1402*	24-28 Jul 44	Lt. Harry E. Taylor, USNR
USS *PCS-1403*	24 Jul-10 Aug 44	Lt. Nelson L. Barnes Jr., USNR
USS *PCS-1452*	24 Jul-10 Aug 44	Lt. (jg) W. B. Norwood, USNR
USS *PCS-1457*	24 Jul-10 Aug 44	Lt. (jg) Frank A. Woodke, USNR
USS *PCS-1461*	24 Jul-10 Aug 44	Lt. Willis S. Harrison, USNR

8

Harbor Stretchers at Tinian

This force will capture, occupy and defend SAIPAN, TINIAN, and GUAM, and initiate the development of advanced bases and airfield on captured islands.

Vessels of Service Squadron TWELVE (Harbor Stretchers) will report to Task Force Commanders for duty as required by them.

—Commander, Joint Expeditionary Force (Vice Adm. Richmond K. Turner, USN) Operation Plan No. A10-44, 6 May 1944.

Service Squadron Twelve, known as the "harbor stretchers," was established in March 1944, for the primary purpose of increasing depths in channels and harbors where major Fleet units would anchor or where coral reefs and shallow water created serious navigation hazards. Its ships and personnel engaged in both removal of navigational hazards, and harbor improvements including the installation of anti-torpedo netting to safeguard ships inside harbors from enemy torpedo attacks. One of the ships assigned duty with Service Squadron Twelve at Tinian was the net laying ship USS *Papaya* (AN-49).[1]

Papaya was built by the Pollock-Stockton Shipbuilding Company, and placed in commission at Stockton on 1 December 1943 with Lt. Comdr. Elias Johnson, USNR, in command. He, along with three other officers, identified below, and fifty enlisted men comprised the ship's company:

- Lt. Richard W. Melum, USNR (Executive Officer)
- Lt. (jg) Henry A. Strickland, USNR (Engineering Officer)
- Boatswain David V. Strubb, USNR (First Lieutenant)[2]

On 8 December, *Papaya* got under way; leaving her builder's yard, she proceeded out the Stockton Deepwater Ship Channel and into the San Joaquin River. Sailing downriver (in a generally westward direction out of California's central valley), she entered Suisun Bay and, continuing on, passed through the Carquinez Strait and into San Pablo Bay, a northern extension of the San Francisco Bay. From there, it was a relatively short jaunt southward to Oakland.[3]

Map 8-1

Water route from Stockton, California, to the Pacific (a distance of about ninety miles)

Arriving at Moore's Drydocking and Shipbuilding Co., Oakland, California, *Papaya* entered drydock for a short fitting out period. Resting on keel blocks, workers coated the ship's bottom with creosote, and installed sonar and radar equipment in other parts of the ship. After departing the drydock, *Papaya* made stops at two USN facilities in the Bay Area. At the Naval Ammunition Depot, Mare Island, she took aboard her allowance of ammunition; and then net and boom gear at the Naval Net Depot, Tiburon, California.[4]

Mare Island is a peninsula in the city of Vallejo, about 23 miles northeast of San Francisco; and Tiburon, a town situated just north of San Francisco, on the far side of the Golden Gate Bridge.

Photo 8-1

The yard net tender USS *Aloe* (YN-1)—later reclassified net laying ship USS *Aloe* (AN-6)—at the net depot at Tiburon, California, circa 1941. The cannister on her foc's'le contains anti-torpedo netting as do those alongside her. Prominent lifting "horns" at the bow of the ship make her easily recognizable as a net layer.
National Archives photograph #80-G-350013

Photo 8-2

Anti-torpedo netting stored in large cannisters at Tiburon, circa April 1941.
Naval History and Heritage photograph #NH 109962

On 31 December 1943, *Papaya*—fitted out and with ammunition and boom and net gear aboard for self-defense and carrying out her mission—reported for shakedown training to commander, Fleet Operational Training Command, Pacific. This training was conducted by Subordinate Command, Treasure Island, California. Situated in the San Francisco Bay, Treasure Island was an artificial island (built in 1936-37 for the 1939 Golden Gate International Exposition) which hosted a Naval Station by this name and other naval facilities.[5]

Photo 8-3

Naval Station Treasure Island, San Francisco, California, circa 1942.
Naval History and Heritage Command photograph #L39-04.08.01

Photo 8-4

Painting by John Falter, circa 1943, of a WAVE with a silhouette of a machine gun. At Treasure Island in San Francisco, WAVES served as gunnery instructors, teaching sailors how to shoot anti-aircraft guns. Women received this specialized training at the Naval Training Station, Great Lakes, Illinois. A common reference to female sailors, WAVE was an acronym for "Women Accepted for Volunteer Emergency Service."
Naval History and Heritage Command photograph #45-127-O

Papaya completed shakedown training on 24 January 1944 and, on 7 February, left the West Coast en route to war duty in the Central Pacific, via a stop at Pearl Harbor. Heavy seas were encountered for the first three days out, which rolled the ship up to 56 degrees and accounted for a majority of her crew becoming "deathly seasick"—this being their first time at sea. Following her arrival at Pearl Harbor, for necessary maintenance, fueling and provisioning, *Papaya* left Hawaiian waters in convoy with other vessels, bound for Majuro Atoll, Marshall Islands, with two tugs in tow.[6]

Arriving at Majuro on 8 March, the net laying ship reported for duty to commander, Service Squadron Ten. Ensuing operations there consisted mainly of laying moorings and channel buoys for the survey ship USS *Bowditch* (AGS-4).[7]

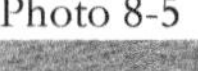
Photo 8-5

Survey ship USS *Bowditch* (AGS-4), circa 1944.
Naval History and Heritage Command photograph #NH 82203

On 4 April, *Papaya* departed Majuro and arrived at Eniwetok Atoll. Her work there was installing and maintaining anti-torpedo nets, laying moorings, and installing channel buoys, as well as numerous miscellaneous jobs. Anti-torpedo nets were intended to block and stop any torpedoes fired from enemy submarines, surface ships, or aircraft. The nets, suspended from floats, were emplaced by net laying ships as a defensive harbor measure to guard shipping at anchor or within a port, and block enemy submarine entry as well.[8]

Photo 8-6

Net layer USS *Sandalwood* (AN-32) tending anti-torpedo nets at Morehead City, North Carolina, in early 1944.
Naval History and Heritage Command photograph #NH 73445

Photo 8-7

Unidentified net layer preparing to drop a buoy from her foc's'le for an anti-submarine net in 1945. USS *Winterberry* (AN-56) is in the background.
National Archives photograph #80-G-327852

ARRIVAL / OPERATIONS IN COMBAT AREA

Departing Eniwetok in convoy on 27 July, *Papaya* arrived at Saipan on 1 August and anchored in Tanapag Harbor. Also reporting for duty to her assigned unit Service Squadron Twelve, were the net cargo ships *Sagittarius* (AKN-2) and *Tuscana* (AKN-3), and net layers *Chinquapin* (AN-17) and *Mimosa* (AN-26). For the next two days, *Papaya* and *Chinquapin* prepared for net laying operations. On 4 August, *Papaya*'s crew completed rigging ship for net operations and performed repairs to an anti-torpedo (A/T) net in the main Saipan Harbor and, on the following day, rigged and laid a net and mooring. That same day, 5 August, *Sagittarius*, *Papaya*, and *Mimosa* received orders to begin Tinian net operations. Tinian Island lay six miles directly SSW of Saipan.[9]

On 6 August, *Papaya* began her part in the installation of an anti-torpedo net baffle in the harbor at Tinian Island. The Allied invasion and occupation of Tinian had commenced on 24 July (designated "J-Day") and fighting on the beach was still in progress when net operations began at the southwest end of the island.[10]

Photo 8-8

Marine LVTs head for the Tinian Landing Beaches, 24 July 1944.
National Archives photograph #26-G-2682

Installation of the anti-torpedo net barrier required the fabrication of net panels with curtain protection by *Sagittarius* and then their transfer, along with associated net moorings to *Papaya* and *Mimosa*. These net layers laid the moorings, installed the net panels, and "buttoned in" these adjacent sections of the anti-torpedo net.[11]

Photo 8-9

Net cargo ship USS *Sagittarius* (AKN-2) with her commanding officer, Comdr. Lyle B. Hillsinger, USNR, standing on the pier, circa August 1945. U.S. Navy photograph

On 21 August, *Sagittarius* completed the fabrication and launching of the twenty-second and final section of the A/T net for Tinian Island. She then transferred her remaining moorings to *Papaya*; discharged temporary embarked personnel and equipment; restowed cargo, and made ready for sea. She left Tinian in the afternoon, and after proceeding to Saipan, anchored for the night. *Papaya* rigged and laid moorings for the final section of the Tinian net, which *Mimosa* buttoned in. Afterward these two net layers also made their way that afternoon to anchorage in Saipan Harbor to prepare to assist in laying nets there.[12]

From 6-21 August, *Sagittarius* had assembled and launched from cargo, 3,000 yards of A/T net for the Tinian Harbor net installation. *Papaya* and *Mimosa*, working in conjunction with the *Sagittarius*, installed all the net sections.[13]

BATTLE STARS

USS *Sagittarius* earned a battle star for her efforts at Saipan as part of the Marianas operation; *Papaya* and *Mimosa* each earned ones for both Saipan and Tinian.

Ship	Saipan	Tinian
Sagittarius (AKN-2)	1-6 Aug 44	
Papaya (AN-49)	1-10 Aug 44	6-15 Aug 44
Mimosa (AN-26)	20 Jun-10 Aug 44	24 Jul-10 Aug 44

9

Consolidation of Northern Solomons

Like everyone else in the Southwest Pacific, I soon found myself falling into the habit of referring to "MacArthur's troops," "MacArthur's planes," and "MacArthur's ships." In the Navy Department in 1942, a friend heard I was going to the Southwest Pacific and wisecracked, "So you are leaving the U.S. Navy to join MacArthur's Navy. God help you. No one else will.

—Vice Adm. Daniel E. Barbey, USN (Ret.), in his memoir, *MacArthur's Amphibious Navy: Seventh Amphibious Force Operations 1943-1945.*

From the summer of 1942 to the summer of 1943, the South Pacific was the most active Pacific theater of operations. Supply lines, for the most part, ran directly from the West Coast of the mainland to bases in Samoa, the Fijis, New Zealand, and New Caledonia, and soon to the New Hebrides Islands and to the Solomons. Supplies also passed through this area en route Australia and the Southwest Pacific area, ruled off on charts as being under General MacArthur.

—Commander Service Force, U.S. Pacific Fleet, History of Service Force – forwarding of, 31 January 1946.

From the beginning of the New Georgia Campaign in June 1943—a component of the larger Solomon Islands Campaign—South Pacific Forces had continuously operated to the west of 159° E. Longitude, a part of the Southwest Pacific Area under the command of Gen. Douglas MacArthur. Ships and aircraft when in this area were under the strategic direction of the Supreme Commander, Allied Forces, Southwest Pacific Area (MacArthur), but were tactically under commander, South Pacific Area and South Pacific Force (Adm. William F. Halsey, Jr., USN). See map on next page.[1]

An explanation of this arrangement is in order. On 22 February 1942, on the eve of the fall of the Philippines to Japanese forces, President Franklin D. Roosevelt had ordered Gen. Douglas MacArthur to proceed from the Philippines to Australia and establish a new headquarters there. At the time, American and Filipino forces were trying desperately to hold the Bataan Peninsula, which formed the northern boundary of the entrance to Manila Bay, and nearby Corregidor Island that divided the opening into two channels. Lt. John D. Bulkeley's Motor Torpedo Boat Squadron 3—*PT-32*, *PT-34*, *PT-35*, and *PT-41*—transported MacArthur as well as his family and members of his staff, and Rear Adm. Francis W. Rockwell and his staff, to Mindanao on 11 March, from whence B-17s flew them to Australia.[2]

On 17 March 1942, two weeks after Gen. Douglas MacArthur's arrival in Australia, the Joint Chiefs of Staff ordered the Pacific divided into two commands, the Pacific Ocean Areas under Adm. Chester W. Nimitz and the Southwest Pacific Area under MacArthur. Nimitz's Pacific Ocean Areas was subdivided into the North, Central, and South Pacific Areas. Although the Solomon Islands west of Guadalcanal and the Russell Islands were in MacArthur's theater, Adm. Ernest King, commander, U.S. Fleet, in Washington, D.C., refused to divide his fleet or to put Admiral Halsey under MacArthur's direct control. This created an awkward command arrangement in which Halsey had operational control of all naval units involved in the Solomons, while MacArthur provided strategic direction.[3]

Map 9-1

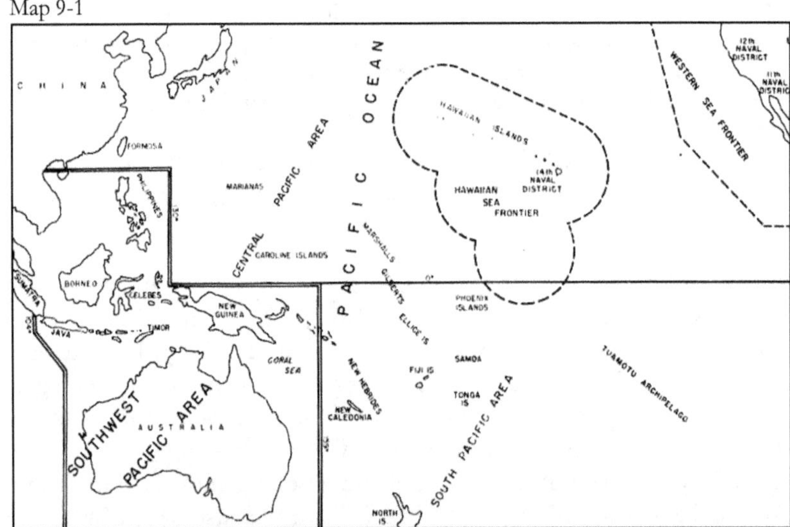

Central Pacific, South Pacific, and Southwest Pacific Command Areas; and Hawaiian Sea Frontier and portion of Western Sea Frontier, 15 April 1944
Commander in Chief, U.S. Pacific Fleet and Pacific Ocean Areas, Operations in Pacific Ocean Areas – April 1944, 23 July 1944

There was also disagreement between the Services over Pacific strategy. MacArthur argued that Allied forces in the Southwest Pacific Area (notably Australia) should undertake an offensive toward the

Philippine Islands. Navy "top brass" called for a major offensive through the island chains of the Central Pacific Area. Ultimately, the question of strategy was resolved through a series of compromises as each theater undertook its own offensive.[4]

Photo 9-1

Gen. Douglas MacArthur (at left) and Adm. Chester W. Nimitz (right) discuss Pacific war strategy at MacArthur's headquarters in Brisbane, Australia, on 27 March 1944. National Archives photograph SC 190409

MACARTHUR AND HALSEY'S PREVIOUS EFFORTS

After defending Port Moresby, Papua, from a Japanese invasion, and establishing a base at Milne Bay, MacArthur's Australian and American forces had advanced northwestward up the east coast of Papua and New Guinea, along his "road" back to the Philippines. Meantime, Halsey's forces moved northwestward up the Solomons to Green Island on the northwest edge of the Solomon Island chain, and Emirau Island in the Bismarck Archipelago (see Map 9-2).

CHANGES IN AREAS OF RESPONSIBILITY

By June 1944, South Pacific Force activities in the Solomons-Bismarck region were entirely confined to blockading—particularly the powerful Japanese naval and air base at Rabaul, on the northeast coast of New Britain. (MacArthur had initially wanted to capture this enemy stronghold, which was a major impediment on his "road back to the Philippines" through the Bismarck Sea. However, mindful of the anticipated large number of Allied combat casualties such action would likely accrue, he had decided, instead, to isolate and bypass Rabaul.) In view of the then minimal role of South Pacific Forces, it was deemed advisable to unify them under the Southwest Pacific Command.[5]

Map 9-2

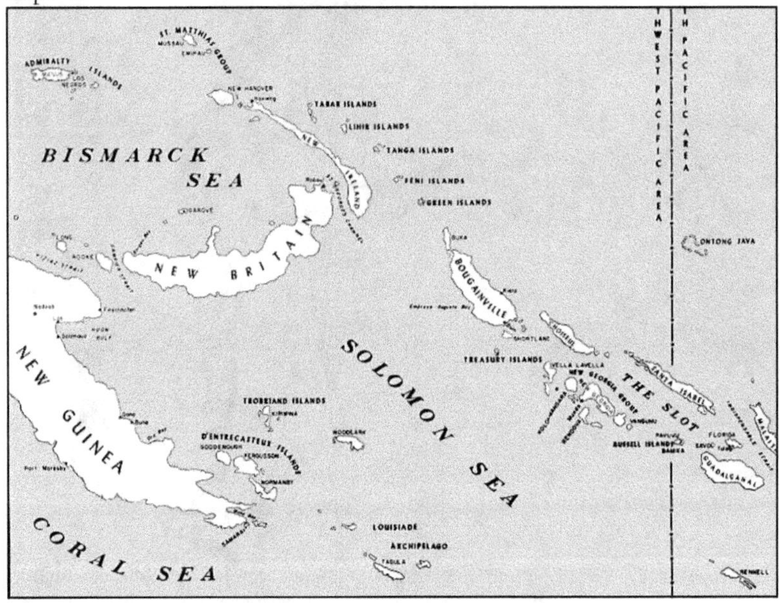

Bismarck and Solomon seas and adjacent island areas
http://ibiblio.org/hyperwar/USMC/II/maps/USMC-II-I.jpg

As a consequence of that decision, at 0100 on 15 June 1944, all military responsibility for the area west of 159° East Longitude and south of the equator, passed to General MacArthur. All aircraft, ships, and troops, as well as bases and installations in this area were transferred at this time.[6]

The South Pacific Force continued its patrol and escort operations, involving responsibility for protecting the vital supply route from the United States to the Southwest Pacific. The logistic support of the bases and areas transferred (except for Emirau in the St. Matthias group) remained under commander, South Pacific, until such time as the Seventh Fleet ("MacArthur's navy") could assume the task.[7]

ADMIRAL HALSEY RELINQUISHES COMMAND OF SOUTH PACIFIC FORCE, TAKES THE HELM OF U.S. THIRD FLEET

> *With my detachment on 15 June 1944 as Commander South Pacific Force and Area, and the simultaneous assumption by General MacArthur of the command of all forces West of Longitude 159° East, the South Pacific campaign against Japanese forces virtually ended.*
>
> —Commander Third Fleet (formerly Commander South Pacific Force and Area), South Pacific Campaign – Narrative account, 3 September 1944.

Photo 9-2

War poster depicting Adm. William Halsey.

Concurrent with the transfer of South Pacific Forces to MacArthur on 15 June, Adm. William F. Halsey Jr., USN, was detached from command of the South Pacific Force. The following day, he assumed command of the Third Fleet, and was also designated commander, Western Pacific Task Forces. Beginning in August 1944, his forces inflicted great losses upon the Japanese fleet in the Palaus, Philippines, Formosa (now Taiwan), at Okinawa, and in the South China Sea.[8]

STOCKTON YMSs JOIN MACARTHUR'S NAVY

Photo 9-3

USS *YMS-97* under way in San Francisco Bay, 1945-1946.
Naval History and Heritage Command photograph #NH 81376

Getting down to the level of individual ships, the units transferred on 15 June 1944 to commander, Seventh Fleet ("MacArthur's navy") included 4 destroyers, 12 infantry landing craft, 7 motor torpedo boat squadrons, 8 motor gunboats, 6 sub-chasers, and 17 auxiliaries. Twenty-three of these ships, which are identified in the table, came from commander, Service Squadron, South Pacific Force. Of those, six were 136-foot minesweepers, which included Colberg Boat Works-built USS *YMS-95* and *YMS-97* from Stockton, California.

Ship	Ship	Ship	Ship	Ship	Ship
YMS-95 ★	*SC-518* ★	*YF-680*	*YO-144*	*YOG-14*	*APc-96*
YMS-97 ★	*SC-722*	*YF-684*	*YO-145*	*YOG-19*	
YMS-196 ★	*SC-723*	*YF-782*		*YOG-40*	
YMS-197 ★	*SC-1267* ★	*YF-783*		*YOG-42*	
YMS-259 ★	*SC-1269* ★				
YMS-269 ★	*SC-1274* ★[9]				

YMS: Yard minesweeper
SC: Sub-chaser
YF: Self-propelled covered lighter
YO: Self-propelled fuel oil barge
YOG: Self-propelled gasoline barge
APc: Small coastal transport
★ Battle star for Consolidation of Northern Solomons

Sister ships *YMS-95* and *YMS-97* each earned a battle star for the consolidation of the Northern Solomons for the period 15 June to 26 October 1944, but written records about their activities are practically non-existent. These and other small ships "disappeared" into "MacArthur's navy."

Consolidation of Northern Solomons

Ship	Commanding Officer	Period
USS *YMS-95*	Lt.(jg) George E. Newby Jr., USNR	15 Jun-26 Oct 44
USS *YMS-97*	Lt.(jg) G. C. Wilkinson, USNR	15 Jun-26 Oct 44

The only references the author could find to these YMSs were contained in the war diaries of other ships, such as those of the auxiliary fleet tug USS *ATA-123* and the motor torpedo boat tender USS *Varuna* (AGP-5). The below entries reveal that on the dates cited, *YMS-97* and *YMS-95* were employed as escort vessels in the Solomon Islands.

> 0745 – Underway [from Lunga Point, Guadalcanal] enroute Munda [New Georgia] with 7 x 3 pontoon tender barge in tow as ordered by CTG [Commander, Task Group] 35.6. USS *YMS-97* as escort vessel. Speed 8 knots.
>
> —USS *ATA-123* 13 June 1944 war diary entry

> 0630 – Underway from anchorage in Nissan Lagoon, Green Island [enroute Blanche Harbor, Treasury Island], by direct authority, in company with USS *YMS-95*.
>
> —USS *Varuna* (AGP-5) 31 July 1944 war diary entry

The Stockton ships may also have engaged in minesweeping during the period for which they earned battle stars, in support of amphibious operations conducted by Rear Adm. Daniel Barbey, USN, commander, Seventh Amphibious Force. In 1944, Barbey's force planned and carried out twenty highly successful operations—removing or neutralizing the Japanese strength in New Guinea, Schouten Islands, Molucca Islands, and a major portion of the Philippines.[10]

SUMMARY OF SOUTHWEST PACIFIC OPERATIONS IN SUMMER AND AUTUMN 1944

In June 1944, activities in MacArthur's Southwest Pacific Theater were largely concentrated on the completion of the Biak Campaign, and the continued blockade and neutralization of enemy forces in New Guinea and the Bismarck Archipelago.[11]

On 2 July, Southwest Pacific forces occupied Noemfoor Island, and the Cape Sansapor area, and on 30-31 July, Vogelkop Peninsula, New Guinea. Throughout the month, consolidation of existing Allied positions, and neutralization of by-passed enemy positions by land, sea, and air forces continued as in previous months.[12]

Map 9-3

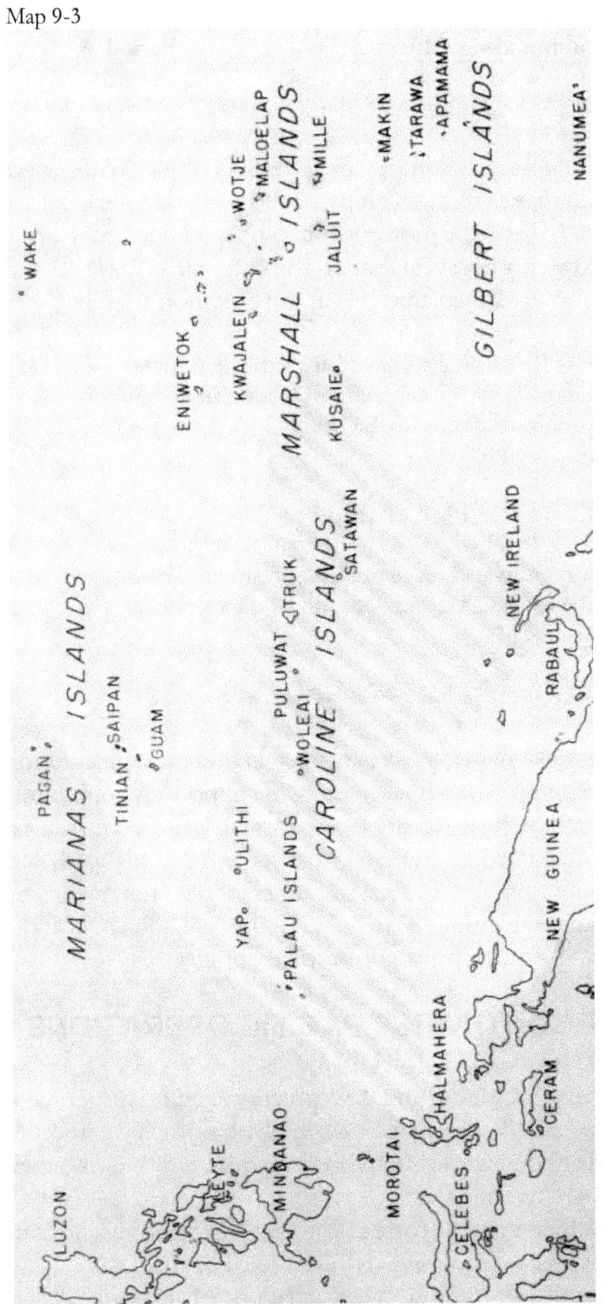

Western Caroline Islands and adjacent areas
Commander in Chief, U.S. Pacific Fleet and Pacific Ocean Areas, Operations in Pacific Ocean Areas – September 1944, 7 March 1945

Operations during August 1944 followed the pattern established in recent months. They were concerned primarily with the consolidation of existing positions from the Northern Solomons to the Cape Sansapor region of New Guinea. Surface and air forces continued to block movement of isolated enemy groups, and supported ground forces in further destruction of the enemy.[13]

On 15 September, the first amphibious landing was made on Morotai Island in the Haimaheras by Southwest Pacific forces. The capture of this island allowed development of airstrips which brought all of Mindanao in the Philippines within the 500-mile radius of land-based aircraft, and allowed longer-ranged aircraft to reach as far as Luzon and Borneo.[14]

Subsequent landings on Morotai were made concurrently with the Central Pacific assault on Peleliu. These former Japanese-held outposts constituted the last barriers remaining before Allied advancement into the Philippines, and constituted utmost value in subsequent assaults on that archipelago.[15]

Activity in the Southwest Pacific theater in October centered on the amphibious landings in the Philippine Islands. Transported and covered (screened) by United States and Australian naval units, Army forces landed at Leyte Island on 20 October. (This operation is the subject of the following chapter.)[16]

The operations in the Central Philippines finally brought together in a significant way, forces of Admiral Nimitz's Pacific Ocean Areas command, and those of General MacArthur's Southwest Pacific Area command. Under Nimitz, Halsey's South Pacific Forces had evolved into the U.S. Third Fleet, and Spruance's Central Pacific Forces into the U.S. Fifth Fleet. U.S. Southwest Pacific naval forces, which operated directly under MacArthur's command had similarly been upgraded to a U.S. Seventh Fleet designation under Vice Adm. Thomas C. Kinkaid, USN.

10

Stockton-built Rescue Tugs

Photo 10-1

Rescue tug USS *ATR-52* under way in San Francisco Bay, 14 August 1944. National Archives photograph #19-N-71102

Colberg Boat Works in Stockton produced four 165-foot, wooden-hulled *ATR-1* class rescue tugs for the Navy. These modestly built, small ships, like all tugs, were designed for power and not speed. Propelled by two Foster Wheeler "D"-type boilers, sending steam to one Fulton Iron Works vertical triple-expansion reciprocating steam engine, driving a single propeller, they had a top speed of 12.2 knots. Ship's armament consisted of one 3"/50 dual purpose gun mount and two single 20mm gun mounts for anti-air defense. Ship's complement was 5 officers and 47 enlisted men.[1]

Colberg Boat Works, Stockton, California, *ATR-1* Class Rescue Tugs

Ship	Comm.	Commanding Officer
USS *ATR-50*	5 May 44	Lt. August B. Billig, USN
USS *ATR-51*	22 Jun 44	Lt. A. L. Larson, USNR
USS *ATR-52*	31 Jul 44	Lt. Charles A. Miller, USN
USS *ATR-53*	11 Sep 44	Lt. B. M. Stevenson, USNR

On 5 May 1944, USS *ATR-50* was accepted and placed in commission at 1420 alongside Colberg Boat Works Outfitting Wharf #3, with the Stockton Field Band assisting in the ceremonies. Lt. (jg) August B. Billig, USN, assumed command of the ship at 1425, with four other officers and fifty men comprising the ship's crew, in attendance. The officers are identified below:

- Ens. George W. Kingston
- Ens. R. A. Dewey
- Chief Boatswain C. J. Edwards
- Machinist H. P. Kimmel[2]

Lieutenant Billig then gave the order, "Set the watch, first section, port routine," establishing the security of the ship. Duty section watch personnel stationed on the quarterdeck began the use of ship's time (obtained from a chronometer, wound daily to maintain spring tension), and made the first entries in the ship's official USN deck log.[3]

SHIP OUTFITTING

The rescue tug remained berthed alongside the wharf until the morning of 9 June. At 1100, with boiler #2 already on line for auxiliary purposes, boiler #1 was lit off. Ship's engineers then tested the safety valves for both boilers and, with these safety devices functioning properly, cut in #1 boiler on the main steam line. The steering gear was tested satisfactorily at 1230.[4]

Fifteen minutes later, pilot William E. Ingham came on board and *ATR-50* got under way at 1310 for the Tiburon Floating Drydock, San Francisco Bay. During the transit, Pilot Ingham had the conn (was giving rudder and engine orders) and the captain and executive officer were on the bridge. Steering various courses and speeds, the rescue tug traversed the Stockton Channel and San Joaquin River. At 1550, she passed Antioch, a city in the East Bay situated along the Sacramento–San Joaquin River Delta, abeam to port. Continuing her westward passage, *ATR-50* left Pittsburg a city on the southern shore of the Suisun Bay in the East Bay region abeam to port at 1620.[5]

Other well-known land marks in the San Francisco Bay were then traversed or passed in succession before arriving at Molate Point to fuel the ship, before continuing on to Tiburon:

- 1735: Passed under Martinez bridge
- 1747: Passed [city of] Benecia, Calif., abeam to starboard
- 1802: Passed under Carquinez bridge
- 1840: Passed Pinole Point abeam to port

- 1910: Moored alongside fueling dock at Molate Point, San Francisco Bay, starboard side to[6]

That evening, having received on board 36 barrels of Navy standard bunker fuel and 5 barrels of diesel fuel, *ATR-50* got under way at 2101 for the anchorage off Chauncey Point, in the San Francisco Bay. At 2124 she "dropped the hook" there, and pilot Ingham left the ship.[7]

Late the following morning, *ATR-50* proceeded at 1100 to the floating drydock USS *ARD-20* a short distance away. She crossed the sill of the drydock at 1116, and by 1210 the dock was pumped down, and she was resting on keel blocks. (There is no information in the ship's log regarding why she was drydocked.)[8]

On 12 May, the rescue tug left dry dock; proceeded to the Molate Point Fueling Dock to top off fuel storage tanks, then to Treasure Island in the San Francisco Bay. Arriving there at 1758, she moored in berth C-3, alongside North Pier, Sea Frontier Base, Treasure Island.[9]

Three days later, a truck from the Mare Island Naval Ammunition Depot delivered the following ammunition to her:

- 150 rounds 3"/50-caliber AA
- 152 rounds 3"/50-caliber AP
- 4,380 rounds 20mm AA
- 8,700 rounds small arms ammunition
- 1 destructor[10]

CALIBRATION, TRAINING AND CERTIFICATION

The remainder of May was devoted to the calibration of equipment and crew training and certification at Treasure Island and San Pedro, California. In waters near Treasure Island, *ATR-50* calibrated her radio direction finder on the Compass and Range Finder range; made runs on the Degaussing Range to calibrate her degaussing equipment (designed to protect the ship against magnetic mines); and compensated (adjusted) her magnetic compasses. The crew was exercised at Fire Quarters (directing streams of firefighting water at the scene of fires), gun drills, and General Quarters (manning battle stations).[11]

The morning of 20 May, a diving air compressor was hoisted aboard. That evening having completed all her preparations in the San Francisco area, *ATR-50* got under way from Treasure Island, bound for San Pedro down the California coast. Standing out of San Francisco Bay, she passed Alcatraz Island abeam to starboard, passed under the Golden Gate Bridge and entered international waters. Departing the main ship channel, ship's course was 212°T, speed 11 knots. Further seaward, after leaving the Farallon Islands to starboard, *ATR-50* came left to 160°T and proceeded southward.[12]

During passage to San Pedro, Billig exercised the crew at Fire Quarters and General Quarters, and test fired the 3-inch gun. At 0420 on the morning of 22 May, radar contact was made on Catalina Island; lights of the San Pedro fishing fleet were sighted at 0506; and sunrise observed at 0603. Approaching Los Angeles Harbor, *ATR-50* passed a sentinel station vessel, then Point Fermin light abeam to port, traversed the swept channel, and entered the harbor. She briefly went alongside the fueling dock at the Target Repair Base to "top off," before moving to an anchorage in the harbor. In mid-afternoon, she proceeded to the Cerritos Channel Navy Dock, and moored alongside Berth 1.[13]

In early afternoon the following day, 23 May, crew training began in San Pedro Bay. It consisted that day of a man-overboard drill, followed by exercising the crew at Fire Quarters, General Quarters, and an abandon-ship drill. On the 24th, while under way in the outer harbor, the crew was again exercised at General Quarters. Next, a full power run over a measured mile to validate the engineering plant was made, followed by more training at Fire and General Quarters.[14]

In ensuing days, other training was also included such as a collision drill, exercising the Fire and Rescue Party, and mooring and anchoring drills. The end of May found *ATR-50* still engaged in training in preparation for war duty in the Asiatic Pacific Theater. *ATR-51*, *ATR-52*, and *ATR-53* would serve there as well, following their outfitting and workups, likely similar to those described in detail for *ATR-50*.[15]

SALVAGE OF USS *CANBERRA* AND USS *HOUSTON*

> *The retirement of the* CANBERRA *and* HOUSTON *from the shores of Formosa while under constant attack by the enemy will become a Navy tradition for skill and guts. A well done to all who assisted in the job.*
>
> —Adm. William F. Halsey Jr., USN, commander, Third Fleet.

The passage of the rescue tug to the war zone is not described but less than six months after her commissioning, *ATR-50* assisted in the salvage operation of the torpedoed cruisers USS *Houston* (CL-81) and USS *Canberra* (CA-70). In October 1944, she and other tugs helped tow the damaged capital ships from Formosa (Taiwan) to safe haven at Ulithi Atoll. United States amphibious forces had captured that atoll in the westernmost area of the Caroline Islands the preceding month, for use by the Navy as a fleet base.

A brief description of the battle in which these vessels were damaged follows. At dawn on 12 October, Task Force 38 (Fast Carrier Groups, Pacific)—a component of Admiral Halsey's Third Fleet—arrived at its launching position, 50-90 miles east of Formosa. Earlier at 0544, a predawn "fighter sweep" to gain command of the air over Formosa and the Pescadores, had been launched. No fewer than 1,378 sorties were flown that day from all four carrier groups for strikes on Formosa airfields and shipping. One-third of Vice Adm. Shigeru Fukudome's Japanese air strength on Formosa was destroyed in aerial combat by the first American strike. Only about sixty of his fighters were operational when the second wave of carrier aircraft came over, and none left the ground to intercept the third. Nevertheless, American Task Force 38 losses were heavy – 48 planes.[16]

That evening, Japanese "Betty" bombers (Mitsubishi G4M Navy Type 1 land-based attack aircraft) armed with torpedoes began a series of raids and harassing operations, which continued until midnight. Fukudome admitted losing 42 planes in these raids with nothing to show for their efforts.[17]

Photo 10-2

Japanese aerial torpedo hits the starboard quarter of the light cruiser USS *Houston* (CL-81) during enemy action on 16 October 1944. *Houston* had been torpedoed amidships on 14 October, while off Formosa, and was under tow by USS *Pawnee* (ATF-74) when enemy torpedo planes hit her again. Heavy cruiser USS *Canberra* (CA-70), also torpedoed off Formosa, is under tow in the distance.
Naval History and Heritage Command photograph #NH 98826

At dawn on 13 October, Task Force 38 arrived at its launching position about seventy miles east of Seikoo Roadstead, Formosa. The pattern of strikes was similar to that of the first day, but fewer sorties were flow, only 974, all before noon. During the evening twilight and as carrier planes were being recovered, low flying Bettys attacked the carrier forces and their escorts, with the result that heavy cruiser USS *Canberra* was hit by an aerial torpedo and badly crippled. The following evening, 14 October, the light cruiser USS *Houston* also took a torpedo hit, and on the 16th, suffered a second one. Subsequent successful efforts to tow these severely damaged ships out of enemy waters was one of the notable salvage exploits of the war.[18]

On 13 October, there being no tugs with Task Force 38, Vice Adm. John S. McCain, commander Task Group 38.1 ordered the heavy cruiser USS *Wichita* (CA-45) to stand by to take *Canberra* under tow. That evening, *Wichita* took strain on the wire towing hawser attached to *Canberra*, and the heavy cruiser was able to make good four knots headway with her tow.[19]

Meanwhile, the fleet tug *Munsee* (ATF-107) was summoned from a fleet oiler group (TG 30.8) standing by at sea, which she and *Pawnee* (ATF-74) providentially, had been ordered to join on 8 October. On the morning of 15 October, *Munsee* arrived and took over care of *Canberra*. Earlier at 0300, the heavy cruiser *Boston* (CA-69) had taken *Houston* in tow. The fleet tug *Pawnee* arrived in late morning on 16 October, and took over tow of *Houston*.[20]

Photo 10-3

USS *Canberra* under tow toward Ulithi Atoll, with USS *Houston* in the right background. The tugs are USS *Munsee* with *Canberra*, and USS *Pawnee* with *Houston*.
Naval History and Heritage Command photograph #NH 98343

Munsee towing *Canberra* and *Pawnee* towing *Houston*, steamed abreast one mile apart as they proceeded slowly onward, screened by Rear Adm. Laurance T. DuBose's Task Group 30.3, while the light carriers *Cabot* (CVL-28) and *Cowpens* (CVL-25) operated in support about 15 to 20 miles away. Task Group 30.3 (Salvage Group) was an ad hoc group quickly assembled to protect *Canberra.* It consisted of DuBose's flagship, light cruiser *Santa Fe* (CL-60), the light cruisers *Birmingham* (CL-62) and *Mobile* (CL-63), heavy cruiser *Wichita* (CA-45), and six destroyers. The mission for Task Group 30.3 was as follows:

> To successfully bring the torpedoed *HOUSTON* and *CANBERRA* from a point deep in enemy waters off FORMOSA to the friendly base at ULITHI, defending the damaged ships against enemy air, surface, or submarine action. Should Task Group 30.3 encounter overwhelmingly superior enemy forces, the damaged ships were to be sunk to prevent their falling into enemy hands, and the remaining units of the Task Group were to retire with a minimum of loss or damage.[21]

The peculiar disposition of Task Group 30.3 presented unique challenges in relative ship movements. The tugs and their ponderous charges could make only 3½ knots speed of advance, while the cruiser and destroyers zigzagged violently at 12 knots in order to maintain their relative positions for air and submarine defense, yet not expose themselves to submarine attack. In simple terms, because of submarine "limiting lines of approach," they were going slower than was safe for them. Opportunity for a submarine to get close enough to a ship to carry out a torpedo attack, greatly increases, the slower the target. Given that a sub might draw close enough, the cruisers and destroyers were zigzagging to hinder the enemy in calculating an accurate firing solution. Like hunting birds on the wing, it was necessary for a sub to lead a target ship when firing a straight-running torpedo. Zigzagging was like a bird dramatically changing its previous flight path before shotgun pellets could reach it, thereby causing a miss and ensuring its own safety.[22]

When an enemy air attack developed or threatened, the screen steamed at 15 knots in a tight circle around the tow.[23]

A day earlier on 15 October, a report was received that a Japanese surface force had departed from the Empire (some Japanese controlled base) to destroy the crippled cruisers. This was followed by a message from commander, Third Fleet, directing commander, Task Group 30.3, to transmit urgent dummy messages during the day. Thus, the predicament of the damaged cruisers was to be turned around. Task

Group 30.3 was to be the bait in a trap set to catch a major portion of the enemy fleet.[24]

Propaganda broadcasts from Tokyo made astonishing claims of achieving smashing victories in the actions off Formosa, and promised further triumphs and the "mopping up" of cripples. Task Group 30.3, now known as the "broken remnants of the Third Fleet" and steaming slowly southeastward, prepared to meet the expected enemy moves.[25]

TASK GROUP 30.3 COMES UNDER AIR ATTACK

In early afternoon on 16 October, two large groups of "bogies" (unknown aircraft) were reported 75 miles westward of the Task Group 30.3. The light carriers *Cabot* and *Cowpens* began scrambling fighters, and intercept was begun by this combat air patrol (CAP). "Tally-ho" (enemy sighted) was received from the CAP, reporting 60-75 planes of which some avoided their attacks and got through. At 1340, Task Group 30.3 came under air attack. The intense fire of all ships of the salvage group, including the tugs and the cripples, resulted in the destruction of two attacking planes. However, the *Houston* was torpedoed for the second time in three days, receiving a hit well aft on her starboard quarter which did considerable damage.[26]

Destroyers from Task Group 30.3, which had been stationed well away from the main body of ships, patrolling in search of the expected enemy surface force, rejoined at 0500 on 17 October. A message from commander, Third Fleet, reported the location of the enemy force some 400 miles to the northeast, hugging the Nansei Shoto island chain. This force was reported to be comprised of 1 carrier, 2 battleships, 1 heavy cruiser, 2 light cruisers, and 8 destroyers. At about 1000, word was received that the American trap had failed. After Japanese search planes sighted Task Force 38, the enemy surface force had lost interest in the mission, and was hastily returning to its base.[27]

TRANSIT CONTINUES AMID STORM WARNINGS

On 18 October, a typhoon was reported to be forming in the South Philippine Sea and moving northwestward along the projected track of the salvage group. Commander, Third Fleet, directed the task group to change course to the south, to clear the path of the typhoon. That day brought long, heavy swells, causing both the damaged ships to roll excessively; hampering damage control operations, particularly aboard the *Houston.* The following day, the 19th, dawned fair, but the heavy swells continued. At 0800, the salvage ship *Current* (ARS-22), under the command of Lt. Comdr. James B. Duffy, Jr., USNR, reported for duty and was directed to assist the *Houston.*[28]

Commander, Third Fleet, ordered the task group to change course at midnight toward Ulithi. Admiral Halsey also directed that a destroyer from Task Group 30.3 be sent forty-five miles to the northwest to rendezvous with Task Group 30.7 (an anti-submarine group consisting of one escort carrier, and four destroyer escorts), the *ATR-50*, and the destroyer *Nicholson* (DD-442) of Task Group 38.4, to direct these units to the location of Task Group 30.3.[29]

RESCUE TUG *ATR-50* JOINS SALVAGE GROUP

On the morning of 20 October, Task Group 30.7 reported for duty, and were directed to take station 10-15 miles ahead of the towing group and conduct anti-submarine operations. Soon afterward, the *Nicholson*; minesweeper *Trever* (DMS-16); and the tugs *Zuni*, *ATR-50*, and *Watch Hill* arrived. *Nicholson* joined the other destroyers, as did the *Trever* (a World War I destroyer converted to minesweeper), which was assigned to the screen of the towing group. By 1130, the fleet tugs *Zuni* and *Pawnee* were towing the *Houston* in tandem, and fleet tug *Munsee* and civilian tug *Watch Hill* were towing the *Canberra*. (Tandem towing is when two tugs are used to pull one tow unit.) The added tugs increased the speed of advance by about 1 knot.[30]

Considerable difficulty was experienced towing the damaged *Houston*, because of her tendency to head into the wind, 20-30 degrees off course. The light cruiser endeavored to correct this by rigging canvas on the bow to increase her sail area (defined as the topside area of a ship acted upon by the wind) forward, and by changing the position of the rudder, which was being operated by hand. These efforts had some effect, but she continued to track badly.[31]

That evening, the towline between the *Pawnee* and the *Zuni* carried away, releasing the *Zuni* which was positioned ahead of *Pawnee*. Although the two ships were directed to resumed tandem towing at daybreak, this did not happen. At 1000 the following morning, 21 May, operations began to transfer the tows of the damaged cruisers from *Pawnee* and *Munsee* to *Zuni* and the other tugs. This involved a long five-hour delay in which formation speed was zero. At 1500, the formation again began to make headway, with *Current* and *Zuni* towing the *Houston*, and *Watch Hill* and *ATR-50* the *Canberra*. Released from their previous duties, *Pawnee* and *Munsee* departed for Peleliu, one of the Palau Islands.[32]

Taking over the tow of *Canberra* involved *Munsee* casting off the tow wire and *Watch Hill* taking it, connecting her to *Canberra*. *ATR-50* then accepted a tow wire from *Watch Hill*, joining her to the other tug, thereby enabling the two of them to tow *Canberra* in tandem. *Watch Hill* was a 186-foot, ocean-going tug built by Globe Shipbuilding of Superior, Wisconsin, in 1943—one of forty-nine War Shipping Administration V4-M-A1-class tugs constructed for the U.S. Navy in various American shipyards for wartime service. She was civilian manned; operated by the Moran Towing Company of New York.[33]

Both were single-screw tugs, but the steel-hulled *Watch Hill* was slightly larger than the 165-foot *ATR-50*, and more powerful—rated at 2,250 hp to the wooden vessel's 1,600 horsepower.

At sunrise on 27 October, Task Group 30.3 arrived off Mugai Channel, Ulithi. At 0545, various tugs stood out to assist the vessels' entry. *Houston* stood (proceeded with towing assistance) toward the entrance to the channel at 0610. Forty-five minutes later, *ATR-50* cast off while *Watch Hill* remained made up as the single towing vessel. At 0927, *Canberra* stood toward the channel, passed through the entrance buoys, then the anti-torpedo net, and proceeded to her anchorage.[34]

AFTERMATH

The effect on Fleet morale of getting damaged ships out of enemy-controlled waters is comparable to the effect that the air-sea rescue service has upon the morale of pilots and aircrewmen. The knowledge that everything will be done to get a severely damaged ship out of a combat area and provide for its safe retirement is bound to bolster the determination of officers and men to stick with their ship.

—Commander, Cruiser Division Ten, Report of Salvage of USS *Canberra* and USS *Houston*, 30 November 1944.

Photo 10-4

Destroyer USS *Killen* (DD-593), at left, and heavy cruiser USS *Canberra* (CA-70), right, undergoing battle damage repairs in the floating drydock *ABSD-2* at Manus, Admiralty Islands, 2 December 1944.
National Archives photograph #80-G-304096

The rescue tug USS *ATR-50* got a taste of war during the tow of USS *Canberra* and USS *Houston* to safe haven at Ulithi Atoll. The following month, November 1944, sister ship *ATR-51* witnessed first-hand devastation wrought by the enemy at Ulithi anchorage—which proved not to be as safe a naval base as was previously believed. (This action is taken up in Chapter 12.)

11

Leyte Landings

At 0559, 18 October 1944, underway from Humboldt Bay, Hollandia, New Guinea to San Pedro Bay, Leyte Gulf, Philippine Islands, with good escort and in convoy including approximately seventy-five auxiliary, amphibious, and merchant ships. Zigzag course not steered.

—USS *Indus* (AKN-1) War Diary entry for 18 October 1944.

Photo 11-1

Net cargo ship USS *Indus* (AKN-1) at anchor, circa February 1944.
National Archives photograph #19-N-62101

In autumn 1944, the U.S. Sixth Army went ashore at Leyte, making good General Douglas MacArthur's vow, "I came through and I shall return." This declaration was made at a train station at Terowie after his arrival by plane in Australia. He, with his family and some staff, had been evacuated on 11 March 1942 from the island of Corregidor, following the Japanese invasion of the Philippines.

As part of the invasion of the Philippines, in early morning on 18 October 1944, the Service Force vessels of Task Unit 77.7.2 (identified on the following page)—less USS *Satinleaf*, USS *Murzim*, and USS

Achilles—stood out of the harbor at Hollandia and took position in a large convoy, bound for San Pedro Bay, Leyte, Philippine Islands, to the northwest. (*Achilles* had departed Hollandia two days earlier, on 16 October, bound for Leyte Island, and *Satinleaf* and *Murzim* would follow on 23 October.) Hollandia was a port on the north coast of New Guinea, located to the west of Wewak.[1]

Map 11-1

Area of Oceania
https://maps.lib.utexas.edu/maps/australia/oceania_pol_2013.pdf

Task Unit 77.7.2: Capt. Emory P. Hylant, USN

Tankers	**Repair and Salvage Ships**
USS *Arethusa* (IX-135)	USS *Achilles* (ARL-41)
USS *Caribou* (IX-114)	USS *Cable* (ARS-19)
USS *Mink* (IX-123)	**Water Ship**
USS *Panda* (IX-125)	USS *Severn* (AO-61)
Underwater Defense Ships	**Ammunition Ship**
USS *Indus* (AKN-1)	USS *Murzim* (AK-95)[2]
USS *Satinleaf* (AN-43)	
USS *Silverbell* (AN-51)	
USS *Teak* (AN-35)	

The convoy departing Hollandia that day was Task Group 78.7 (Reinforcement Group Two) with Task Unit 77.7.2 ships in company. Totaling over seventy ships, the convoy was composed of tank landing ships, merchant vessels, screening ships, and the task unit vessels. The task unit was to remain with the convoy until the approaches to San

Pedro Bay were reached, at which time the Service Force ships would be released for operational control to Capt. Emory P. Hylant, USN, who was embarked aboard the 441-foot net cargo ship USS *Indus* (AKN-1). Hylant was the commander, Service Force, Seventh Fleet, coordinator for the Leyte operation. He and his staff had reported aboard *Indus* at Milne Bay (located at the eastern tip of Papua New Guinea) while she was loading net gear there from 28 September to 4 October 1944.[3]

Map 11-2

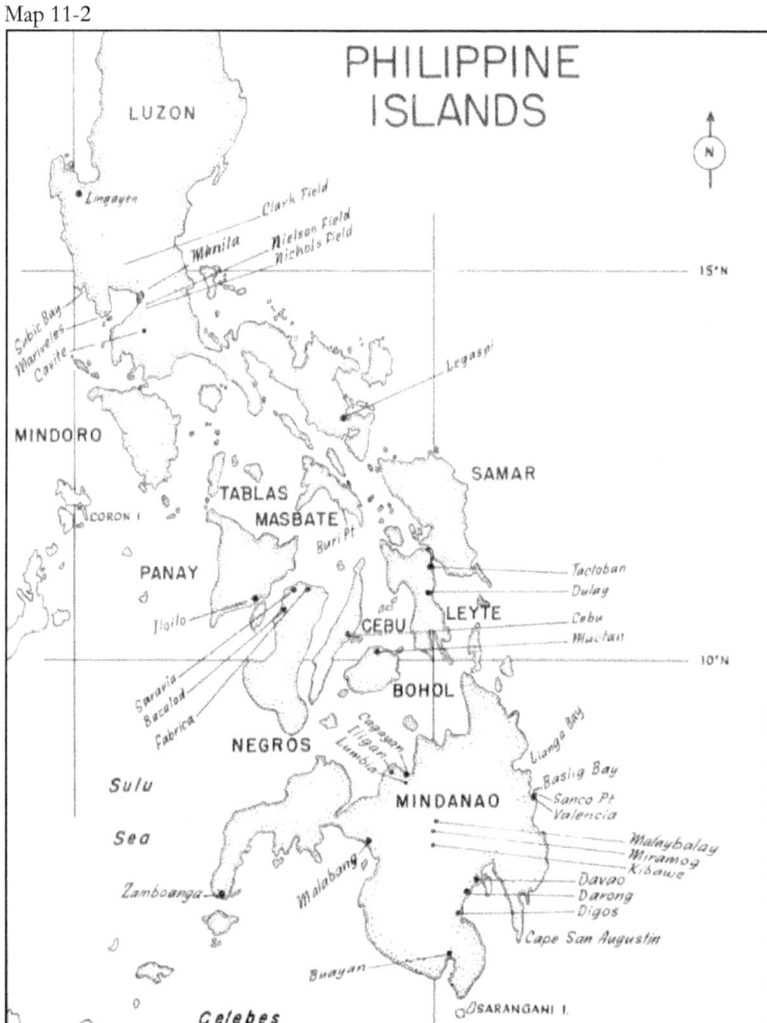

Philippine Islands
Commander in Chief, U.S. Pacific Fleet and Pacific Ocean Areas, Operations in the Pacific Ocean Areas - September 1944, 7 March 1945

Upon release of the Task Unit 77.7.2 Service Forces ships from the convoy, nearing the approaches to San Pedro Bay at the northwest end of Leyte Gulf, Hylant was to assume command of the task unit and confirm this action by hoisting the flag signal AFFIRM MIKE BAKER aboard the *Indus*. He was then to form the ships into a column with *Indus* leading. Upon arrival in the bay, the mission of the task unit was to provide logistic support to the Central Philippines Attack Force.[4]

The voyage to Leyte Gulf was uneventful, aside from loss of a vessel and sighting of a mine. At 1000 on 18 October, the first morning out from Hollandia, the tanker USS *Arethusa* (IX-135)—capable of about 10 knots on a good day—was taken in tow by the rescue and salvage ship USS *Cable* in an effort to maintain convoy speed. That afternoon, she was detached and directed to return to port independently. *Arethusa*'s IX designation signified that she was an unclassified miscellaneous vessel, a category the Navy used for ships that did not fit neatly into established ones. *Arethusa* was the former tanker *Gargoyle*, built in 1921 at Oakland, California, by the Moore Shipbuilding Co. She had been acquired by the Navy on 23 March 1944 from the War Shipping Administration on a bareboat basis for use as a mobile floating storage tanker. (The term "bareboat" refers to the chartering or hiring of a ship or boat, whereby no crew or provisions are included as part of the agreement.)[5]

The remaining convoy units proceeded without incident until the evening of 20 October when, at 1734, the patrol frigate *Muskogee* (PF-49) sighted a mine which was sunk by destroyer *Hopewell* (DD-681) with gunfire. During this interlude, movement of the convoy toward Leyte continued unabated.[6]

At 1930 on 23 October, the convoy front was reduced to 2,000 yards in width to facilitate entrance into Leyte Gulf (via a narrow channel off the south coast of Homonhon Island), and lessen the danger from reported mines. The following day, the tank landing ships in company were detached to proceed to Yellow Beach Two, White Beach, and Red Beach; and Task Unit 77.7.2 to the harbor in San Pedro Bay; after which the convoy was dissolved. Destroyer Squadron 21, ships previously assigned convoy screening duty, then established a patrol off Capiness Point, Leyte Gulf, awaiting further instructions.[7]

ALLIED LANDINGS AT LEYTE ON 20 OCTOBER

Should we lose in the Philippines operations, even though the fleet should be left, the shipping lane to the south would be completely cut off so that the fleet, if it

> *should come back to Japanese waters, could not obtain its fuel supply. If it should remain in southern waters, it could not receive supplies of ammunition and arms. There would be no sense in saving the fleet at the expense of the loss of the Philippines.*
>
> —Adm. Soemu Toyoda, Imperial Japanese Navy, discussing Vice Adm. Takeo Kurita's mission to destroy completely the transports in Leyte Bay following the American invasion of the Philippines, and why there were no restrictions as to the damage that his force might sustain.[8]

The U.S. Sixth Army went ashore at Leyte Island, on 20 October 1944, two months and two years after the first landings were conducted in the Guadalcanal area of the Solomon Islands. This significant operation was predicated and made possible by the ones that opened the way. Amphibious landings in the Marianas, in June and July 1944, breached Japan's strategic inner defense ring and gave the Americans a base from which B29 bombers could attack the Japanese home islands. Following the invasion of Saipan, the Japanese counterattacked in the Battle of the Philippine Sea, fought between the First Mobile Fleet and American Fifth Fleet, from 19 to 21 June. In what one American aviator termed "The Great Marianas Turkey Shoot," the U.S. Navy destroyed three enemy aircraft carriers—the *Hijo*, *Shokaku*, and *Taiho*—some 480 planes, and nearly as many aviators. The devastating loss left the Japanese with virtually no carrier-based aircraft or experienced pilots for the forthcoming Battle of Leyte Gulf.[9]

MACARTHUR AND HALSEY JOIN FORCES

The naval force (comprised of units of the American Third and Seventh Fleets) assembled for the invasion of Leyte was not quite as large as the one that had taken part in June in the invasion of Normandy, but it had more striking power. Embarked aboard the assault vessels were the U.S. Sixth Army's Tenth and Fourteenth Corps commanded by Lt. Gen. Walter Krueger, USA.[10]

Per direction by Nimitz, forces previously led by Halsey and Spruance were essentially combined. When Halsey commanded them, they were designated the Third Fleet, and when Spruance, alternating with Halsey, took the reins, the Fifth Fleet. Following the landings by Central Pacific (Third Fleet) forces at Palau, and those by Southwest Pacific Forces on Morotai, planned landings on Yap (phase II of the Palau operation) by Third Fleet forces were cancelled, and the forces that were to carry them out were assigned to the Seventh Fleet for the landings at Leyte.[11]

General MacArthur was overall in charge of the Philippines operation and, under him, Vice Admiral Kinkaid, the naval operations. The three major task organizations of the Seventh Fleet were:

- Task Force 77 (Covering and Support) under Vice Adm. Thomas C. Kinkaid, USN (as well as overall naval operations)
- Task Force 78 (Northern Attack Force) under Rear Adm. Daniel E. Barbey, USN
- Task Force 79 (Southern Attack Force) under Vice Adm. Theodore S. Wilkinson, USN[12]

Covering the landing operations, though remaining under the control of commander in chief, Pacific Fleet and Pacific Ocean Areas (Admiral Nimitz), were the fast battleships and carriers of the Third Fleet, commanded by Admiral Halsey.[13]

The weather at the entrance to Leyte Gulf at daybreak on 20 October was cloudy with altostratus and partial swelling cumulus, a visibility to seaward of twelve miles, and light winds from the southeast. Planners had been concerned that a typhoon might pass through the area and cause retirement or diversion of the forces en route from New Guinea. However, the conditions on "A-Day" (Assault Day) were perfect as described by commander, Third Amphibious Force:

> The assault proceeded on schedule following the preliminary bombardment by ships' gunfire and aircraft, a slight onshore tendency of the almost imperceptible wind conveniently drifting the smoke and dust of the bombardment off the beaches and into the interior. The airborne beach observer had made his required report earlier, but the report was unnecessary in this case due to the almost complete absence of surf.[14]

The landings at Tacloban, located in northeast Leyte on an inlet of the Leyte Gulf, and at Dulag, twenty-five miles to the south, were made against little opposition. Naval historian Samuel Eliot Morison noted about the operation: "The Leyte landings were easy, compared with most amphibious operations in World War II—perfect weather, no surf, no mines or underwater obstacles, slight enemy resistance, mostly mortar fire." With this beginning, the liberation of the Philippines was off to a good start.[15]

SUPPORT BY SERVICE FORCE, SEVENTH FLEET

Service Force, Seventh Fleet, commanded by Rear Adm. Robert O. Glover, USN—and of which the Task Unit 77.7.2 ships were a part—was a small organization. Accordingly, it was ill prepared to provide necessary logistic support for the Leyte operation when the Seventh Fleet was suddenly expanded by dozens of warships lent by the Pacific Fleet. This task was only accomplished by improvisation, stretching supply lines to the breaking point, and the utmost efforts on the part of Glover's officers and men.[16]

For a considerable time after the landings at Leyte, and as troops fought inland, a system of resupply echelons was operated by Service Force, Seventh Fleet. From a pool of five ammunition ships, three cargo ships, three fleet oilers, four auxiliary oilers and a number of gasoline and diesel tankers, an echelon was included in each regular follow-up convoy. The Royal Australian Navy as part of Task Group 77.7 (The Leyte Gulf Service Group of the Seventh Fleet), contributed the ammunition ships HMAS *Yunnan* and *Poyang*, the provision ship HMAS *Merkur*, and the oiler HMAS *Bishopdale*. Groups of these Allied ships in rotation departed Hollandia or Biak Island, Papua New Guinea, or Manus in the Admiralty Islands, every six days for Leyte Gulf.[17]

Photo 11-2

Merchant vessel *Yunnan* at Fremantle, Western Australia, on 24 December 1941. She was requisitioned by the RAN, on 22 June 1942, for service as the ammunition stores-issuing ship HMAS *Yunnan*.
Australian War Memorial photograph 301779

Another supply problem, not previously mentioned was drinking water. Like the other supplies, it was also handled by the Seventh Fleet. It had always been a difficult problem in amphibious operations to supply fresh water for small craft which had no distilling apparatus. In previous operations, large vessels had given of their plenty to the small ones, but in an operation employing such large numbers of small ships and craft, the assistance of the larger ships could not be depended upon. Accordingly, in the summer of 1944 the Navy converted several fleet oilers into fresh water tankers, each of which could carry four million gallons. The *Severn* (an oiler converted to water ship) arrived at Leyte with Task Unit 77.7.2 on 24 October.[18]

ACTIVITIES OF UNDERWATER DEFENSE SHIPS

Photo 11-3

Net laying ship USS *Silverbell* (AN-51) in San Francisco Bay, 3 April 1944.
Naval History and Heritage Command photograph #NH 73441

Net Cargo Ship and Net Laying Ships at Leyte

Ship	Commanding Officer	Battle Star
Indus (AKN-1)	Lt. Comdr. Andreas S. Einmo, USNR	18 Oct-29 Nov 44
Satinleaf (AN-43)	Lt. Arthur B. Church, USNR	
Silverbell (AN-51)	Lt. Harold N. Berg, (D)M, USNR	18 Oct-29 Nov 44
Teak (AN-35)	Lt. Byron Hollett, USNR	

Among the Service Force ships present at Leyte to provide logistic support, were four devoted to helping to safeguard fleet units in San Pedro Bay from attack by Japanese submarines. These were the net cargo ship USS *Indus*, and the net laying ships USS *Satinleaf*, *Silverbell*, and *Teak*. *Silverbell* was a Stockton ship. A modestly equipped and armed product of Pollock-Stockton Shipbuilding Co., she was commissioned on 16 February 1944, Lt. Harold N. Berg, USNR, in command. Propelled by two Busch-Sulzer BS-539 diesel-electric engines, coupled to a single propeller, the 194-foot net layer had a top speed of 12 knots; and was fitted with one single 3"/50 dual-purpose gun mount and four twin 20mm AA gun mounts for self-defense. The ship's complement was equally modest, four officers and fifty-two men.[19]

Information about *Silverbell* is scant. She earned one battle star in the war, at Leyte, and Lt. Berg was presented with a Bronze Star Medal by commander, Service Squadron Four, in June 1945. Little is recorded about her service. This was not uncommon regarding small vessels; many didn't maintain war diaries or, if they did, relied on larger ships to compile and submit information about operations. Nevertheless, we can surmise much about her duties at Leyte, by recalling those of other net layers, and assume that she, like *Indus*, earned her bronze star for action against attacking Japanese aircraft.[20]

In preparation for operations at Leyte, *Satinleaf* was at Brisbane, Australia, from 6-10 October 1944 taking aboard sonar buoy (sono-buoy) material. Departing Brisbane, she proceeded to Milne Bay, New Guinea. After arriving there on the 15th, she embarked personnel of a sonar buoy unit and sailed for Hollandia the next day. Following her arrival at Leyte on 29 October, she was employed for salvage operations and general utility until 5 November. That day she commenced sonar buoy operations, laying and guarding the submarine detection devices until 17 January 1945.[21]

From 17-20 January, *Satinleaf* was employed for general utility duties. On 20 January, she took up a new assignment as the "Harbor Entrance Control Post of San Pedro Bay," which she continued until 12 March 1945.[22]

Photo 11-4

USS *Teak* (AN-35) at San Francisco, California, 21 February 1944.
Naval History and Heritage Command photograph #NH 73435

Teak's service was similar. After arriving in Leyte Gulf on 24 October, for the next few weeks despite frequent calls to General Quarters, she performed general utility duties. These included laying net moorings and marker buoys in the gulf, aiding grounded small craft, and making tows. In late November, she took up sonar buoy station duties between Samar and Homonhon islands. This lasted until 17 January 1945, when she returned to tending and laying moorings.[23]

JAPANESE AIR ATTACKS AT LEYTE

> *During the stay at Leyte, the ship's force issued provisions, fuel, stores, and other miscellaneous gear to small craft, repair boats, and assisted the Staff in its mission. Air raids were numerous and the USS* INDUS *(AKN-1) is credited with having shot down two Japanese planes.*
>
> —From the ship's history of USS *Indus.* Capt. Emory P. Hylant and his Task Unit 77.7.2 staff, left the *Indus* on 29 November 1944.[24]

Photo 11-5

At left: at "Val" (Aichi D3A Navy type 99 carrier bomber) and right, a "Betty" (Mitsubishi G4M Navy Type 1 land-based attack aircraft). Japanese Operational Aircraft CinCPOA 105-45 "Know Your Enemy!"

When the net cargo ship USS *Indus* arrived in San Pedro Bay, Leyte Gulf, on 24 October, her cargo consisted of approximately two nautical miles of anti-torpedo net complete with necessary mooring gear, 180 tons of dry stores together with sufficient clothing and small stores for 500 men for 10 days, and ship's stores sufficient for 3,000 men for a like period.[25]

After *Indus* anchored in the bay in late afternoon on the 24th, there was considerable enemy air activity throughout the remainder of the day and that night. In that period her crew was at General Quarters for a total of 9 hours and 45 minutes, during which one or more of her guns probably assisted in destroying an enemy plane. Claiming a "probable assist" meant that her own ship's gunfire probably hit an enemy plane under fire by other ships as well. A "sure assist" assessment signified confidence that her own ship's guns had scored a hit(s) on an enemy aircraft, assisting in destroying it.

Before progressing into a brief overview of the underwater defense activities undertaken at Leyte, it's illustrative to provide, in the following table, a summary of the time *Indus'* crew spent at their battle stations in late October and war diary entries regarding anti-aircraft fire they observed and provided.

USS *Indus* Time Spent at General Quarters in late October 1944

Date	Time at GQ	Comments
24 Oct	9 hours, 45 minutes	At 1845, commenced firing on enemy aircraft passing overhead; probable assist on one plane destroyed; attack on shipping and beach facilities by "Val" and "Betty" type planes continued
25 Oct	12 hours, 43 minutes	Enemy air activity continued throughout day and night; sure assist with 3"/50 on Japanese medium bomber; various enemy aircraft observed shot down during the day
26 Oct	7 hours, 24 minutes	Japanese suicide plane observed to dive on SS *Benjamin I. Wheeler*; several planes observed destroyed
27 Oct	3 hours, 38 minutes	Sent 33 cans of fomite (a fire-fighting agent) to *Benjamin I. Wheeler* still burning
28 Oct	2 hours, 48 minutes	Enemy air activity decreased
29 Oct	3 hours, 52 minutes	No description available
30 Oct	53 minutes	No description available
31 Oct	2 hours, 18 minutes	Decreased enemy air activity; several enemy planes observed shot down[26]

November was also marked with enemy air activity.

Perhaps no Allied vessel took more punishment at Leyte than the Liberty ship SS *Benjamin Ide Wheeler*. She was defended there for 76 days as her Naval Armed Guards went to General Quarters 353 times and underwent 93 enemy bombing or suicide attacks. Her guns destroyed a plane on 24 October, hit two on 25 October, and, even after being seriously damaged on 27 October, destroyed planes on 29 October and 3 November, and hit a plane that crashed on 14 November. On Christmas Day, she moved to the inner harbor at Tacloban, a port city located on the northeastern tip of Leyte Island. On 7 January 1945 her Armed Guards were taken off.[27]

Aboard *Benjamin Ide Wheeler*, one Armed Guard and one merchant seaman were killed in the 27 October suicide attack that disabled the ship. An *Indus* war diary entry for the 27th lamented, "Assembling, launching, and installation of net defenses apparently postponed indefinitely." However, that day she "loaded AN [an unidentified net laying ship] with mooring gear for channel markers to be laid." The following day, *Indus* "loaded mooring gear aboard AN and barges for installation of various pontoon barges and avgas (aviation gasoline) barge moorings."[28]

Indus continued discharging mooring gear on 29 and 31 October. On 1 November, she "transferred mooring gear to an AN for sono-buoys." Three days later, on 4 October, the net cargo ship's gunfire brought down an enemy plane in early morning darkness:

> At 0330, enemy plane observed coming in low off port bow. Opened fire with 20mm's when plane abeam to port. Range – 1000 yards. Expended 153 rds. 20mm ammunition. Plane shot down by 20-6 [designation of 20mm gun] aboard USS *Indus* and observed to crash in flames two miles off starboard quarter. Enemy aircraft activity continued throughout day.[29]

Later that day, she loaded an AN with six moorings for PT boats.[30]

This pattern of operation continued in November 1944, with periodic enemy air activity, intermixed with loading of sono-buoys and moorings aboard one or more unidentified ANs for underwater defense installations.[31]

12

USS *Mississinewa* Lost to Kaiten Attack at Ulithi

When a midget submarine attack badly damaged Mississinewa *(AO 59) and set her afire on 20 November 1944,* Munsee, Menominee, Extractor *(ARS 15), and* ATR-51 *did excellent fire-fighting work, going alongside the ship with its burning cargo, mainly of aviation gasoline. The work was to no avail, however, and* Mississinewa *sank.*

—Charles A. Bartholomew and William I. Milwee Jr., *Mud, Muscle, and Miracles: Marine Salvage in the United States Navy*, 2d ed., 2009.[1]

Photo 12-1

Fleet oiler USS *Mississinewa* (AO-59) capsized and sinking at the base of the flames, caused by burning aviation fuel on the water's surface, Ulithi Atoll, 20 November 1944. National Archives photograph #80-G-K-5510

On 20 November 1944, the fleet oiler USS *Mississinewa* (AO-59) was sunk by a Japanese manned Kaiten torpedo at Ulithi. The loss of the tanker marked the first successful attack by this new type of suicide weapon—which was part of a larger Kaiten mission that day against U.S. naval forces present at Ulithi Atoll.[2]

Mississinewa was a new ship, only six months old. She had been commissioned at her builder's yard, Bethlehem-Sparrows Point Shipyard, Inc., Sparrows Point, Maryland, on 18 May 1944, Capt. Philip G. Beek, USNR, in command.[3]

Photo 12-2

USS *Mississinewa* (AO-59) under way, circa May 1944. She is painted in camouflage scheme Measure 32, Design 3AO.
Naval History and Heritage Command photograph #NH 97280

Completing shakedown in Chesapeake Bay, she sailed down the U.S. East Coast to Aruba, Netherland West Indies, to load her first cargo. After filling her fuel tanks 23-24 June, she took departure of Caribbean waters, bound for Pearl Harbor via the Panama Canal.[3]

Mississinewa arrived in Pearl Harbor on 10 July, and left there on the 15th for Eniwetok, Marshall Islands. Following her arrival on 24 July, she remained at Eniwetok fueling fleet units until her departure on 20 August for Manus Island, in the Admiralty Group off New Guinea. The fleet oiler reached there on 24 August. On 1 September, she got under way to rendezvous with, and refuel, units of Admiral Halsey's Task Force 38. *Mississinewa*'s real mission had now started, and she was continually on the move while supporting the Palau campaign, the Formosa raids, and the Leyte campaign.[4]

The fleet oiler arrived at the Ulithi Atoll in the Western Caroline Islands, Palau Group, on 15 November 1944. She had aboard, upon arrival, 404,000 gallons of aviation fuel and 9,000 barrels of diesel fuel. After *Mississinewa* anchored in berth #131, a merchant tanker came alongside her, and she began to load Navy Special Fuel Oil from it. Receipt of fuel oil was completed the next day, and *Mississinewa* reported to commander, Service Squadron Ten, "Ready for sea with a standard load."[5]

JAPANESE SUICIDE TORPEDO ATTACK

> *On the morning of the 20th of November 1944, reveille was held at 0530 as usual. I was in my bunk, just about ready to get up, at about 0545, when there was a heavy explosion, which appeared to be on the port side forward, followed immediately by a second very heavy explosion. I was thrown out of my bunk against the after bulkhead of my cabin and landed lying on deck. Somehow or other I knew we had been torpedoed. On looking up I saw flames shooting in through all the port holes similar to a giant blow torch. To get up meant burning to death, so I crawled out into the passageway and stood up. It was very hot.*
>
> *There was a man lying on deck in the passageway apparently unconscious or dead. I dragged this man out on deck and down the ladder to the midship boat deck. There were two enlisted men running aft. I told them to take this man aft with them, put a life jacket on him and throw him over the stern and to pass the word to all hands to abandon ship over the stern.*
>
> —Capt. Philip G. Beek, USNR, describing his immediate actions after the blast from the explosion of a torpedo warhead made clear that his ship was doomed.[6]

Diagram 12-1

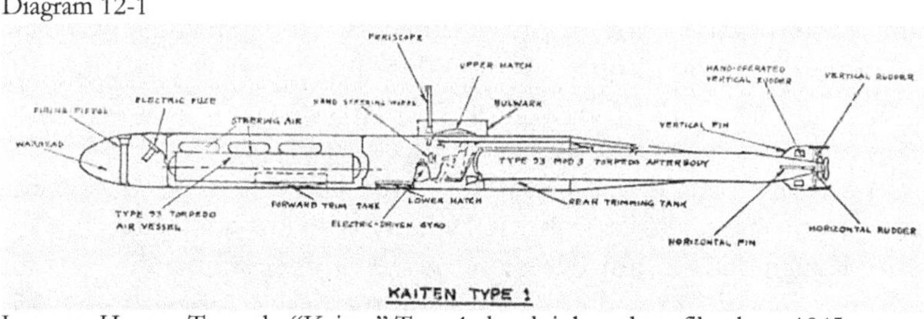

Japanese Human Torpedo "Kaiten" Type 1 sketch inboard profile plans, 1945.
Naval History and Heritage Command photograph #NH 78668

Mississinewa had been attacked by a Type 1 Kaiten in the first of this type mission. The enemy suicide weapon was essentially a Type 93 oxygen-fueled torpedo with a bigger warhead and, for the pilot, a compartment that included a small periscope, steering control, controls to arm and detonate the weapon, batteries, and air filter. The Type 1 had a 3,420-pound warhead, a top speed of 30 knots, a maximum range of 42 nautical miles at a cruising speed of 12 knots, and a maximum operating depth of 250 feet.[7]

As the Japanese became increasingly desperate in early 1944, the idea of manned torpedoes which had been previously rejected was then adopted. Actual development of the Kaiten began in February 1944, and their first operational use came nine months later. On 8 November, the Kikusui-tai Kaiten Group left Otsujima, Japan. The submarines *I-47*, *I-36*, and *I-37* each piggybacked four Kaiten, and each were also armed with eight conventional torpedoes. The plan called for *I-47* and *I-36* to attack the Ulithi Atoll anchorage with their Kaiten, while *I-37* struck a different anchorage near Palau. After launching the suicide Kaiten, the submarines were to proceed to Leyte Gulf to carry out conventional attacks on U.S. shipping supporting MacArthur's ongoing operations on Leyte.[8]

I-47 and *I-36* arrived off Ulithi in early morning on 19 November. At dawn, *I-47* approached to within 4.5 nautical miles of the atoll and reported seeing over 200 U.S. ships in the anchorage inside the reef. Beginning at about 0300 the following morning, *I-47* launched all four Kaiten. The first to be released was piloted by Lt. (jg) Sekio Nishina (the co-inventor of the Kaiten torpedo), believed to be the Kaiten that hit *Mississinewa.* The tiny human-operated torpedo was sighted heading toward the *Mississinewa* by alert lookouts on the fleet oilers USS *Cache* (AO-67) and USS *Lackawanna* (AO-40), but they could take no action, other than to report the enemy, since both ships were at anchor.[9]

Destroyer USS *Case* (DD-370), on patrol at the entrance to the atoll, rammed and sank one of the other Kaitens. Of the remaining two from *I-47*, one ran aground on the reef. The other *Kaiten* made it into the lagoon, and was lost, probably sunk by depth charges from destroyers and destroyer escorts reacting to the attack on *Mississinewa.*[10]

Only one Kaiten was successfully launched from *I-36*. It was not heard from again, and probably fell victim to an attacking aircraft. The other Kaitens would not release from the mother submarine, and the remaining one developed a heavy leak in the pilot's compartment. *I-36* surfaced to recover the pilot in the flooded Kaiten, and was quickly attacked by two Avenger torpedo bombers on anti-submarine warfare patrol; she crash-dived without damage.[11]

EFFORTS TO SAVE *MISSISSINEWA*

Small craft and tugs not suitable for fighting fire continuously embarrassed our position and prevented maneuvering of this vessel to most favorable positions. This vessel is primarily designed and equipped for firefighting.

—Lt. A. L. Larson, USNR, commanding officer, USS *ATR-51*,
a Stockton ship constructed by Colberg Boat Works.[12]

Photo 12-3

USS *ATR-51* in center of photograph with smoke from the burning USS *Mississinewa* (AO-59), apparently capsized at this point, off the rescue tug's starboard side. Commander Service Squadron Ten, Torpedoing of USS *Mississinewa* and events in connection therewith, 8 December 1944

At 0550 on the morning of 20 November, Lieutenant Larson, commanding officer of the rescue tug *ATR-51*, received a report from the bridge that "an apparent explosion and fire existed on vessel in the vicinity of berth 130." He immediately ordered preparations for the tug to get under way, and to fight fire. At 0601, only eleven minutes later, *ATR-51* was proceeding at full speed (185 shaft RPMs) toward the burning ship lying at anchor inside the lagoon.[13]

ATR-51 arrived at the burning *Mississinewa* at 0644 and commenced fighting fire. Approaching from the windward side, she put her bow against the starboard quarter of the burning tanker, and remained there until fire was out in that vicinity of the ship. The rescue tug then shifted position to the port bow of the *Mississinewa*, and continue fighting fire. After the fire on the bow was apparently extinguished, she shifted to midships. As the tanker began to sink, *ATR-51* backed clear and stood by for any necessary further actions.[14]

Photo 12-4

USS *ATR-51* in San Francisco Bay, 1946.
Naval History and Heritage Command photograph #NH 81039

Fire broke out again amidships just before *Mississinewa* sank, and ignited fuel on the water freed from ruptured tanks. The oil continued to burn after the tanker slipped below the surface, until extinguished by the fleet tugs *Munsee* (ATF-107), *Menominee* (ATF-73), and *ATR-51*.[15]

RESCUE OF SURVIVORS FROM *MISSISSINEWA*

> *Insofar as anyone can account for time under such conditions, it is my opinion that we left the ship about 15 minutes after the first two explosions.*
>
> *From the deck of the* CACHE, *I saw my ship completely enveloped in flames over 100 feet high. Tugs arrived at the scene and were pumping furiously on the flames, but to no avail. At about 1000 the ship slowly turned over and disappeared from sight.*

> *Considering the suddenness of the explosions and the intense fire it is nothing less than a miracle that more lives were not lost. Out of a total complement of 298 there were three officers and fifty-seven enlisted personnel lost.*
>
> *I was taken aboard the* CACHE *with several other men, where we received medical attention, food and clothes. Most of us were naked or had our clothes burned off.*
>
> —Capt. Philip G. Beek, USNR.[16]

As fires raged aboard the tanker, and while *ATR-51* was still en route to her, Capt. Philip G. Beek and five other men either jumped overboard, or went down a ladder hanging over the stern into the water. They were the last to leave the ship, survivors of the explosions and resultant flames consuming *Mississinewa*. There was a channel about fifteen feet wide not covered with burning oil; the men made their escape along it. Boats from the *Cache* and *Lackawanna* picked them up.[17]

DESTRUCTION OF KAITENS

Photo 12-5

A destroyer escort dropping depth charges while searching for Japanese midget submarines in Ulithi anchorage on 20 November 1944, following the sinking of the fleet oiler USS *Mississinewa* (AO-59).
National Archives photograph #80-G-270650

At about 0540 that morning, a task force had been leaving Ulithi Harbor, requiring the opening of the anti-torpedo net at the entrance of the atoll for the task force to go out. This provided opportunity for the Kaiten that attacked *Mississinewa* to enter. Following its attack on the tanker, it was seen proceeding further into the lagoon.[18]

In the meantime, the alarm had gone off throughout the harbor, and destroyers closed in on the Kaiten, depth-charged it, and blew it up. Earlier, as the task force was leaving Ulithi, the periscope of a different Kaiten had been sighted, and the destroyer USS *Cassin* (DD-372) had rammed it, sinking the midget submarine. A third Kaiten was bombed by an aircraft about five miles eastward of the entrance.[19]

Photo 12-6

Japanese "Kaiten" Type 1 recovered by U.S. forces at Ulithi Atoll in 1945. This is the after half of the Kaiten. The forward portion, including warhead, forward oxygen and fuel tanks and the crew compartment, is missing, and may have been destroyed in an explosion of the warhead.
National Archives photograph #80-G-350027

13

Assault and Occupation of Iwo Jima

At great cost, you'd take a hill to find then the same enemy suddenly on your flank or rear. The Japanese were not on Iwo Jima. They were in it! I'd known combat in the Solomons with its sly ambushes and jungle firefights, but Iwo was another kind of war. On Iwo by the 8th day, only two officers of my second battalion (26th Marines, 5th Marine Division) were standing.... We had one prisoner—unconscious, his clothes blown off.

—Col. Thomas M. Fields, USMC (Ret.).

The Battles of Iwo Jima and Okinawa were fought by Allied forces between February and August 1945 to secure island bases for a final B29 bomber assault on Japan. The Iwo Jima operation was conducted first because it was expected to be easier than an assault on Okinawa. As a result of the enemy's prolonged and bitter defense of Leyte and Luzon, the planned dates for both actions had slipped from those originally planned. Thus, the Pacific Fleet, originally scheduled to cover just Okinawa, had to cover and support both invasions. Meanwhile the Seventh Fleet and its amphibious forces were concurrently engaged in liberating the Southern Philippines.[1]

On 19 February, as naval gunfire pounded Iwo Jima, more than 450 ships massed off the island. Marines of the 4th and 5th Divisions hit the four assault beaches shortly after 0900, initially finding little enemy resistance. Coarse volcanic sand hampered their movement as they struggled to move up the beach from the surf zone. As the protective naval gunfire subsided to allow for advancement, the Japanese emerged from fortified underground positions to begin a heavy barrage against the invading force. The 4th Marines continued to push forward against heavy opposition to take the Quarry, a Japanese strong point, while the 5th Marine Division's 28th Marines isolated Mount Suribachi that same day.[2]

The 3rd Marine Division joined the fighting on the fifth day, charged with securing the center sector of the island. The fortified enemy defenses linked miles of interlocking caves, concrete blockhouses and pillboxes, requiring frontal assaults to gain nearly every inch of ground. Maj. Gen. Harry Schmidt, commanding the Fifth Amphibious Corps—of which the 3rd, 4th, and 5th Marines were a part—declared Iwo Jima secured on 16 March. Ground fighting, however, continued between then and the official completion of the operation on 26 March 1945.[3]

Map 13-1

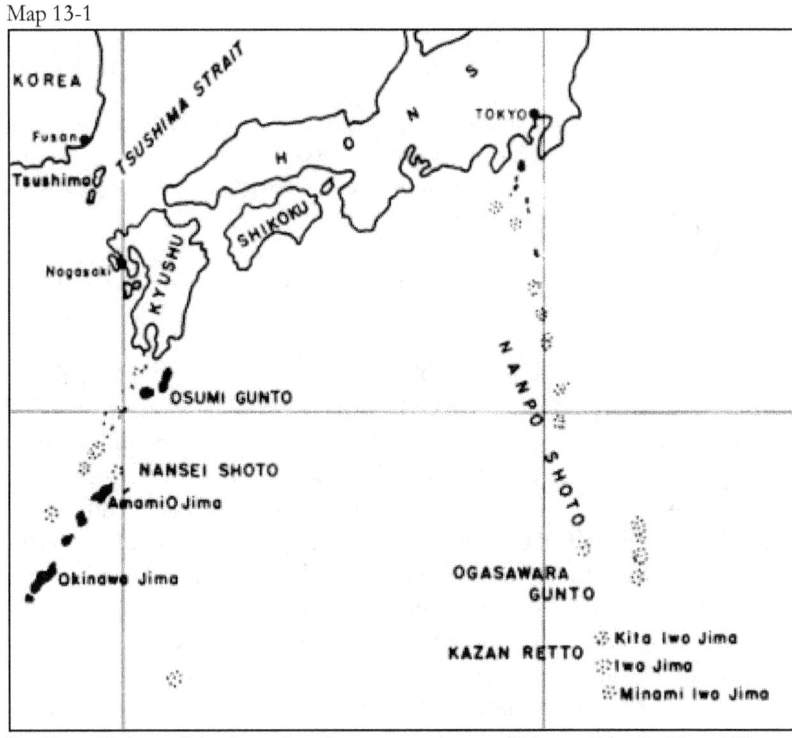

Iwo Jima, Volcano Islands, and to the west-northwest, Okinawa Islands, Nansei Shoto https://www.history.navy.mil/content/history/nhhc/research/library/online-reading-room/title-list-alphabetically/b/building-the-navys-bases/buidling-navys-bases-vol-2-chapter-26.html

COLBERG BOAT WORKS-BUILT SHIPS AT IWO JIMA

Among the armada of U.S. Navy ships at Iwo Jima were four wooden-hulled vessels constructed by Colberg Boat Works in Stockton, California. These were the patrol craft sweepers *PCS-1403* and *PCS-1404*, and the rescue tugs *ATR-51* and *ATR-52*.

The patrol craft sweepers were assigned as control ships, which involved their being on the line of departure during amphibious landings, and guiding combat troop-laden assault craft to the beach.

However, because *PCS-1403* had arrived at Iwo Jima with her radar inoperative owing to a failed component, she was not part of the assault and was assigned alternative duties during the island invasion.

The rescue tugs operated with the Service and Salvage Group (Task Group 51.3) under salvage expert Capt. Lebbeus H. Curtis V, USNR. Curtis flew his broad pennant from the salvage ship USS *Clamp* (ARS-33), commanded by his son, Lt. Comdr. Lebbeus H. Curtis VI, USNR.

ATTACK FORCE AND CONTROL GROUP

Comprising the Attack Force (Task Force 53) for the invasion of Iwo Jima were the ships which carried troops, equipment, and supplies, for putting ashore during the assault phase. Their total was not less than 30 APA (amphibious attack transports), 12 AKA (amphibious cargo ships), 3 LSD (dock landings ships), 1 LSV (vehicle landing ship), 46 LST (tank landing ships), and 30 LSM (medium landing ships), plus associated flagships and escorts.[4]

PCS-1403 was one of the units of the Control Group 53.5, under Capt. Bruce Byron Adell, USN. It controlled the assembly, formation, and movement of scheduled assault waves and other ship-to-shore traffic, as well as other services. (Adell was the control officer on the staff of commander, Amphibious Forces, U.S. Pacific Fleet.)

Control Group (Task Group 53.5): Capt. Bruce Byron Adell, USN
Central Control Unit (Task Unit 53.5.1)
PCE-877
PCS-1403, PCS-1421
Control Unit Able (Task Unit 53.5.2)
PCS-1460, PCS-1461
Control Green (Task Unit 53.5.3)
PC-463, SC-1315
Control Red (Task Unit 53.5.4)
PC-469, SC-1316
Control Unit Baker (Task Unit 53.5.5)
PCS-1452, PCS-1455
SC-1066, SC-1272, SC-1360
Control Yellow (Task Unit 53.5.6)
PC-578
Control Blue (Task Unit 53.5.7)
PC-1081, SC-1374[5]

AMPHIBIOUS CRAFT LANDINGS

> *During morning of invasion, we had a good view of the landing by our forces on the beaches. Landing appeared difficult due to surf conditions, enemy mortar fire*

was heavy, our vehicles appeared to have a difficult time in moving through the sand on the beach.

—USS *PCS(H)-1388* War Diary, February 1945. The converted patrol craft sweeper and sister ship *PCS(H)-1396*, were assigned hydrographic survey ship duties.

Photo 13-1

First waves of the assault force, embarked in LVTs, head for the Iwo Jima beach on D-Day, 19 February 1945. Support landing craft (LCSs), positioned near shore, are providing close fire support; as are, farther offshore, two destroyers and at right the battleship *Tennessee* (BB-43). Mount Suribachi is in the upper left.
Naval History and Heritage Command photograph #NH 104127

Captain Adell was embarked aboard the escort patrol craft *PCE-877*. As shown in the preceding table, the single escort patrol craft (PCE) serving as flagship, as well as six patrol craft sweepers (PCS), six sub-chasers (SC), and four patrol craft (PC) were assigned duties as control ships. Available descriptions of the activities of some provide context for those of the Control Group as a whole.

At 0540 on 19 February, after arriving at Iwo Jima, *PCE-877* secured from her screening duties, proceeded to the transport area to embark, then to the line of approach, and assumed her Central Control Vessel station. Upon arrival there, she began dispatching waves of landing craft per the planned schedule. The escort patrol craft maintained her station, except for a period while standing by as rescue vessel for *LSM-211*. Over the course of the day, *PCE-877* came under fire several times, but received no hits.[6]

Mundane supporting duties followed until 21 February when, on orders of the Force Control Officer, Captain Adell, *PCE-877* left station to conduct rescue and salvage searches and operations, aiding stray and sinking LVTs and LCVPs after poor weather conditions caused them to drift seaward. These rescue and salvage patrols continued until the 24th, when *PCE-877* was ordered to take up screening duties in stations designated C21 and C27. On 26 February, per orders from commander LST Flotilla 21, she joined a departing convoy and took up screening duties.[7]

ACTIVITIES OF OTHER CONTROL VESSELS

On the morning of 19 February, *PCS-1452* arrived on her control vessel station at 0730. With H-Hour set for 0900, the first wave formed up in fine shape, with everything running accordingly to the pre-arranged schedule. During this and the following day, the line of departure and the LST area were subject to sporadic enemy fire. The subject patrol craft sweeper experienced some near misses from enemy high velocity projectiles, believed to be 47mm, but no casualties.[8]

An enemy air raid occurred mid-afternoon on 22 February, with bombs dropped about 3,000 yards to the west of where *PCS-1452* was stationed, but no shipping was hit. She was on station the following evening when, at about 2030, a warning came about a pending air attack. While the PCS proceeded to windward to lay smoke to protect the anchored vessels in her charge, about twelve bombs fell off her port side, fairly close, but again no casualties were suffered aboard. However, an LCM was hit, and the amphibious force command ship *Auburn* (AGC-10) was straddled by dropping bombs.[9]

Photo 13-2

USS *Auburn* (AGC-10) off San Francisco, California, circa November 1945. Naval History and Heritage Command photograph #NH 77383

Two days later, *PCS-1452* again witnessed bomb dropping, this time near the beach, with damage unknown. From 25 to 27 February, she moved in closer to the beach and remained there, controlling movement of supplies and personnel. During this period, she received some small caliber fire from the beach, but nothing significant. There were no air raids on these days.[10]

In early afternoon on 28 February, *PCS-1452* joined a task group as a member of the screen and begin the transit back to Saipan.[11]

Two of the patrol craft sweepers assigned as control vessels—*PCS-1455* and *PCS-1403*—hosted dignitaries, and/or senior and mid-grade military officers and ferried them around the combat area. From 19 to 25 February, *PCS-1455* served as staff vessel for Brig. Gen. Franklin A. Hart, USMC (assistant division commander, Fourth Marine Division). This duty involved maneuvering the ship as necessary, on the right flank of the line of departure, to enable Hart to observe the progress of troops ashore and movement of necessary supplies.[12]

PATROL CRAFT SWEEPER *PCS-1403*

As earlier mentioned, Stockton-built *PCS-1403* arrived off Iwo Jima on 19 February with an inoperative SC-1 radar, owing to a component failure two days earlier. Repair efforts by two crewmembers, including about five hours spent atop the mast in heavy seas and high winds, proved fruitless. At 0500 that morning unable to perform her duties, she was detached from Task Group 51.5 (with which she had transited from Saipan), and closed (moved toward) the amphibious force command ship USS *Eldorado* (AGC-11). The PCS spent the day dodging light caliber fire from the beach, with orders to remain in the area and make smoke for LSTs during the night.[13]

At dawn the following day, 20 February, *PCS-1403* again moved in closer to the flagship, and spent the afternoon towing shipping skids (left by unloading of LCTs) clear of the transport area, and assisting small craft adrift in the vicinity. Like *PCS-1452*, she also experienced some near misses from projectiles, and an air raid, as described in a war diary entry:

> We were bracketed several times during the day by shore fire. Probably spent AA with a very high trajectory. Otherwise, we would have taken several hits, for the shells, which detonated on striking the water, were falling in steps about 30 yards apart.
>
> The usual air raid during the night, but no bombs fell in our area.[14]

On 21-22 February, during separate actions, the patrol craft sweeper ferried Marine officers from the flagship USS *Eldorado* (AGC-11) and heavy cruiser *Indianapolis* (CA-35) to and from Red Beach 1. On one of the return crossings to their ships, officers returned with captured Japanese documents.[15]

Photo 13-3

Secretary of the Navy James Forrestal, Lt. Gen Holland McTyeire ("Howlin' Mad") Smith, USMC, Rear Adm. Harry W. Hill, USN, and Vice Adm. Richmond Kelly Turner, USN, aboard USS *Eldorado* (AGC-11) during Iwo Jima campaign. Photo is signed by James Forrestal: "To a competent, fighting Navy man, Vice-Admiral Kelly Turner." Naval History and Heritage Command photograph

During these two days, *PCS-1403* also took a deserted LVT in tow, but the towline parted during an air raid as she was maneuvering and making smoke. Unable to later locate the tracked landing vehicle (Buffalo), it was presumed to have sunk. Better luck was had with a disabled LCM found adrift. She was able to tow it back to the transport area, and to its parent APA (attack transport).[16]

In late afternoon on 22 February, Lt. John M. Cherry Jr., USNR, *PCS-1403*'s commanding officer, had to secure the port main engine because of excessive vibration, caused by fouling of the screw and shaft, by the broken towline the previous night. Motor Machinist's Mate Second Class Clint H. Norman and Seaman First Class Raymond L. Williams volunteered to dive over the side and attempt to free the propeller in spite of the cold and heavy seas. Using a homemade diving helmet made from a Japanese gas mask and breathing compressed air from the engine room, they were able to free the line. Working in shifts, both men were underwater for about 40 minutes each.[17]

At 0910 on 23 February, the patrol craft sweeper closed the *Eldorado* and embarked the Secretary of the Navy, James Forrestal; Lt. Gen. Holland M. Smith, USMC; Rear Adm. Louis E. Denfield, USN; Rear Adm. Earl W. Mills, USN; and members of their staff for an inspection of the beaches. The party returned to the flagship at 1300.[18]

That evening, while laying to about 3,000 yards off Green Beach, a violent explosion occurred at 1850 about 100 yards off *PCS-1403*'s starboard beam, shaking the ship severely but causing no damage. An Air Flash Red warning was announced over the radio twenty minutes later. It was thought the explosion was caused by a rocket bomb from the beach, or an aerial bomb from an aircraft that had approached undetected.[19]

Over the next three days, the patrol craft sweeper ferried groups of officers to and from Red Beach 1, including on 26 February Lt. General Smith and several members of his staff. That evening, *PCS-1403* received orders at 1850 to proceed to Area William and join Task Unit 51.16.6 as an escort, for return to Saipan. The commander of the task unit of "cripples" was the captain of the salvage ship *Gear* (ARS-34), towing the *LSM-202*, *LCI-473*, and *SC-1027*. Other ships in company were the *LST-716*, *LST-578*, *LST-390*, and *LCI-651*; under escort by the destroyer *Williamson* (DD-244), *PCS-1403*, and the sub-chasers *PC-578* and *SC-1390*.[20]

HEAVY CASUALTIES SUFFERED AT IWO JIMA

The acquisition of Iwo Jima—strategically important as an air base for fighters escorting B29s flying long-range bombing missions against mainland Japan from the Mariana Islands, and as an emergency landing strip for crippled B29s unable to make it back to their base—came at a high cost. The vital link in the chain of bases was gained through the individual and collective courage of Marines over a thirty-six-day period. The brutal fighting resulted in 26,000 American casualties, including

6,800 dead. Only 1,083 of the 20,000 Japanese defenders of Iwo Jima survived.[21]

Photo 13-4

Stars and Stripes flying over Iwo Jima; U.S. Marines raised the American flag over Mount Suribachi, Iwo Jima, on 23 February 1945.
Naval History and Heritage Command photograph #NH 104279

PATROL CRAFT SWEEPER BATTLE STARS

Ship	Assault and Occupation of Iwo Jima Award Period	Commanding Officer
USS *PCS(H)-1388*	19-28 Feb 45	Lt. (jg) Edwin J. Richards, USNR
USS *PCS(H)-1396*	19-28 Feb 45	Lt. Frederic Eugen Sturmer, USNR
USS *PCS-1403*	19-26 Feb 45	Lt. John M. Cherry Jr., USNR
USS *PCS-1404*	20-26 Feb 45	Lt. William H. Beatty Jr., USNR, and/or Lt. (jg) John G. Carlson, USNR
USS *PCS-1421*	19-26 Feb 45	Lt. Edmond T. Freeman, USNR
USS *PCS-1452*	19-28 Feb 45	Lt. (jg) J. S. Simms, USNR
USS *PCS-1455*	19-26 Feb 45	Lt. Carter F. Cort, USNR
USS *PCS-1457*	20-26 Feb 45	Lt. Frank A. Woodke, USNR
USS *PCS-1460*	19 Feb-8 Mar 45	Lt. Foster William Lamb, USNR
USS *PCS-1461*	19-27 Feb 45	Lt. Willis S. Harrison, USNR

SERVICE AND SALVAGE GROUP (CTG 51.3)

The salvage forces for the invasion of Iwo Jima, under the command of Captain Lebbeus Curtis V, USN, consisted of a landing craft repair ship, a battle damage ship, three salvage ships, two fleet tugs, and two rescue tugs, as well as several large support landing craft [LCS(L)3] configured for salvage, and a tank landing ship (LST) fitted with the Brodie aircraft landing system. The two ATR rescue tugs, *ATR-51* and *ATR-52* as previously mentioned were Stockton built. These ships and craft all which earned battle stars are identified in the table on the next page, as well as the periods for which the stars for the Assault and Occupation of Iwo Jima were earned.

A specially adapted tank landing ship *LST-776*—a unit of the Service Component of the Service and Salvage Group—was the only U.S. Navy ship configured with the Brodie system to see combat in World War II. Because her capability was so unique, an explanation of her precedes an overview of the salvage efforts at, and in support of Iwo Jima.[22]

Photo 13-5

USS *LST-776* during Brodie system trials off New Orleans, Louisiana, 1943. U.S. Navy photograph

Ten hours into the invasion of Iwo Jima, *LST-776* launched three Piper Cubs to spot artillery fire for the Marines. One aircraft crashed during takeoff, but the other planes accomplished their missions. The capability arose from U.S. Army Air Force captain James Brodie's innovation in determining how to launch and recover small airplanes from a ship without a runway or a landing deck.[23]

Brodie's Aircraft Landing System allowed a small, single-engine plane like an L-4 Piper Cub or L-5 Sentinel to take off from and land on a non-aviation ship. This was done by utilizing a trolley running along an elevated cable stretched between specially rigged vessel masts. After a hoist and sling device lifted the plane and its crew and slung it below the trolley, a special winch pulled the aircraft back to the ship's stern. The pilot then gunned the plane's engine. When it had gained a high rpm, a clutch released the trolley, and the plane rolled along the cable, picking up speed. As the aircraft accelerated and reached near the limit of travel of the trolley, the pilot pulled on a lanyard which detached the plane from the sling, allowing it to soar into the air.[24]

To land the aircraft, the pilot lined the aircraft up with the cable, and snagged the sling with a hook mounted above its fuselage. The machine's winch worked like a fishing reel, and slowed the plane to a stop. Both takeoffs and landings required only 600 feet of cable—and often less with strong headwinds.[25]

Service and Salvage Group (Task Group 51.3):
Capt. Lebbeus H. Curtis V., USNR

Landing Craft Repair Ship	Battle Star Award Period
Agenor (ARL-3)	19 Feb-16 Mar 45
Battle Damage Repair Ship	
Oceanus (ARB-2)	20 Feb-16 Mar 45
Salvage Ships	
Clamp (ARS-33)	19 Feb-2 Mar 45
Gear (ARS-34)	19-26 Feb 45
Shackle (ARS-9)	19 Feb-10 Mar 45
Fleet Tugs	
Tawakoni (ATF-114)	19 Feb-10 Mar 45
Yuma (ATF-94)	19-26 Feb 45
Rescue Tugs	
ATR-51	18-25 Feb 45
ATR-52	19 Feb-16 Mar 45
Tank Landing Craft	
LST-776	20 Feb-1 Mar 45
Large Support Landing Craft	
LCS(L)3-31	19 Feb-8 Mar 45
LCS(L)3-32	19 Feb-26 Mar 45
LCS(L)3-33	19-26 Feb 45
LCS(L)3-34	19-26 Feb 45
LCS(L)3-35	19-26 Feb 45
LCS(L)3-36	19-26 Feb 45[26]

Navy Captain Lebbeus Curtis was eminently qualified to lead the Service and Salvage Group. One of his most noteworthy projects before the war was the building of the harbor in Port Hueneme, California. In 1940, while then an experienced civilian salvor (salvage expert), he had been hired under a special contract to relieve Comdr. William A. Sullivan at the San Diego salvage base. Sullivan, who had established the salvage station on the U.S. Pacific Coast, had been ordered to England to survey the salvage situation and report how the British were coping with the salvage problems of the war.[27]

When the Japanese attacked Pearl Harbor in 1941 Curtis was a lieutenant commander in the Naval Reserve, and happened to be present in Hawaii on 6 December, en route to the Middle East for salvage work. Awakened by the bombing on the 7th, he hurried to Pearl Harbor. With mammoth work needed to clear the harbor of sunken and damaged ships, unexploded ordnance, and other debris, further travel was unthinkable and he had to remain in Pearl Harbor for several months.[28]

As soon as he could be spared from Pearl Harbor salvage work, Curtis was designated Mobile Salvage Engineer on the staff of Rear Admiral William L. Calhoun, commander, Service Force. Salvage in the central Pacific was then in the hands of Curtis, now a full commander, and he ranged over it, taking his considerable expertise and experience wherever needed. By the time of the invasion of Iwo Jima, he had been promoted again, from Commander to Captain, in recognition of his tremendous achievements.[29]

As previous mentioned, Captain Curtis flew his command pennant at Iwo Jima from USS *Clamp*, the salvage ship commanded by his son, aboard which he was embarked. The salvage forces on the scene were not static, but changed as ships under his command came and went. Curtis had support from backup forces of Service Squadrons Ten and Twelve in the Marshalls and Marianas. As tugs from these squadrons carrying out their assignments supporting combat operations, those arriving at Iwo Jima reported to Task Group 51.3 for duty.[30]

The bulk of the work by salvage ships and tugs at Iwo Jima involved refloating landing craft and ships broached and swamped on the beach. Soft volcanic sand, not passable by amphibious tracked vehicles, created an unusually large amount of such work. Vehicles bogged down in the first waves blocked the beaches, and prevented following waves of landing vessels from either beaching securely or anchoring effectively in the poor holding ground.[31]

Information about the activities of *ATR-51* is scant, but they were likely similar to those of *ATR-52*. The latter rescue tug had departed Apra Harbor, Guam, on 16 February 1945, en route to Iwo Jima as part of Task Group 51.5. Upon arrival there on the 20th, she reported to Task Group 51.3. From 21 February through 8 March, *ATR-52* engaged in salvage operations towing and clearing Iwo Jima beaches of wreckage as part of the salvage group. She departed the island on the 9th as part of Task Unit 51.23.2.[32]

14

Zamboanga Operation

0921: Observed enemy fire falling in vicinity of HMAS Warrego *and USS* Cinnamon, *engaged in hydrographic work on SANTA CRUZ BANK. Apparently 3" or mortar fire from vicinity of TARGET #18. Manned all Battle Stations.*
0931: Commenced counter-battery fire with main and secondary batteries.

—Light cruiser USS *Phoenix* (CL-46) Action Report of Operations Supporting Zamboanga, Philippines, Landing, 3-13 March 1945, entries about an RAN sloop and a USN net laying ship taking enemy shore fire on 9 March off Zamboanga, Mindanao, the day preceding a planned amphibious assault there.

Photo 14-1

Loaded LCVPs ("Higgins boats") head for assault beaches at Zamboanga, Mindanao, during landings there on 10 March 1945.
Naval History and Heritage Command photograph #80-G-308949

In late winter/early spring 1945, following the Allied liberation of Leyte, and as that of Luzon continued, Gen. Douglas MacArthur began efforts to expel the Japanese from the entire Philippine Archipelago utilizing the U.S. Seventh Fleet and Eighth Army—which were determined not to be urgently required for the invasions of Iwo Jima and Okinawa. The resultant series of amphibious assault operations in the Southern Philippines and Borneo are identified in the table:

Operations in the Southern Philippines and Borneo, 1945

Area	Code Name	D-day
Palawan	VICTOR III	28 February 1945
Zamboanga	VICTOR IV	10 March 1945
Panay and W. Negros	VICTOR I	18 March 1945
Cebu	VICTOR II	26 March 1945
Bohol	VICTOR II	11 April 1945
S.E. Negros	VICTOR II	26 April 1945
Mindanao	VICTOR V	17 April 1945
Tarakan, Borneo	OBOE I	1 May 1945
Brunei Bay, Borneo	OBOE VI	10 June 1945
Balikpapan, Borneo	OBOE II	1 July 1945[1]

By the beginning of March 1945, combined operations of Allied ground, naval, and air forces on Luzon—where the Japanese had concentrated the greater portion of their Philippines defense garrisons—had split the enemy into five separate isolated areas. The Japanese were well dug in for defense in northern Luzon, in the mountains east of Manila, and near the port area of Manila; but were no longer capable of coordinated action. Forthcoming Allied effort in the remainder of the Philippines would involve gradual elimination of isolated enemy garrisons, rather than any single major engagement.[2]

An aspect of the second of these operations—the landings on Zamboanga Peninsula, Mindanao—involved the Royal Australian Navy sloop HMAS *Warrego* (U73) and net laying ship USS *Cinnamon* (AN-50), performing pre-assault survey work under fire. Commanded at the time by Lt. Comdr. George D. Tancred, RAN, *Warrego* gained many Battle Honours while undertaking extensive service during the war:

- PACIFIC 1941-45
- DARWIN 1942-43
- NEW GUINEA 1942-44
- LINGAYEN GULF 1945
- BORNEO 1945

Photo 14-2

HMAS *Warrego* at Port Moresby, Papua, September 1942.
Australian War Memorial photograph 026608

After being commissioned into the RAN on 22 August 1940, HMAS *Warrego* performed a variety of duties—basically whatever she was asked to do—including patrol and escort, minesweeping, anti-submarine and, at this junction, survey ship duties. In July 1944, she was assigned to the U.S. Seventh Fleet Survey Group (TG 70.5) for duties in New Guinea waters. Prior to this assignment, in early 1943, the RAN surveying ships were formed into Task Group 70.5, part of the U.S. 7th Fleet, while the included Australian Hydrographic Service was designated the charting authority for Allied Naval Forces in the Southwest Pacific.[3]

The newer USS *Cinnamon*, built at the Pollock-Stockton Shipyard, Stockton, California, was commissioned on 10 January 1944. She was smaller than *Warrego*, at 194 feet in length, being three-quarters that of the 266-foot sloop. Lt. Comdr. James H. Russell, USNR, was *Cinnamon*'s first commanding officer, and remained serving in that capacity throughout the duration of the war.[4]

Photo 14-3

USS *Cinnamon* (AN-50), date and location unknown.
Courtesy of NavSource and Gene Maresca

OPERATION VICTOR IV

The purpose of the landings at Zamboanga (10 March) and the following ones at Basilan Island (16 March) was to establish control over areas adjacent to the Basilan Straits; destroy enemy forces therein; and establish facilities to support future operations to reoccupy the southern Philippine Archipelago.[5]

Map 14-1

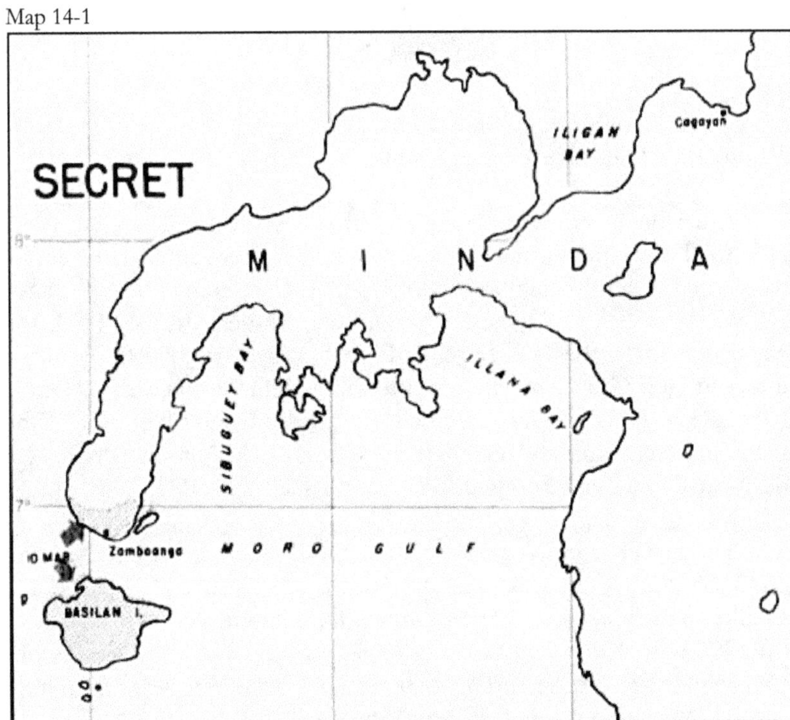

Southwest Mindanao; landings at Zamboanga and on Basilan Island, across the Basilan Strait to the immediate south, identified by arrows at lower left
Commander in Chief, U.S. Pacific Fleet and Pacific Ocean Areas, Operations in Pacific Ocean Areas – March 1945, 31 August 1945

NAVAL FORCES ASSIGNED

The Zamboanga Attack Group (Task Group 78.1), under the command of Rear Adm. Forrest B. Royal, USN, was charged with transporting and landing the U.S. Army 41st Infantry Division, Reinforced (less the 186th Regimental Combat Team). The composition of the Attack Group follows:

- Flagship *Rocky Mount* (AGC-3)
- Cruiser Covering Group (Task Group 74.3): Cruisers *Phoenix* (CL-46) and *Boise* (CL-47), and destroyers *Taylor* (DD-468), *Nicholas* (DD-449), *O'Bannon* (DD-450), *Fletcher* (DD-445), *Jenkins* (DD-447), and *Abbott* (DD-629)
- Motor Torpedo Boat Unit (Task Unit 70.1.11): Tender *Oyster Bay* (AGP-6) and MTB Squadrons 8 and 24 (21 PT boats)
- Minesweeping Unit (Task Unit 78.1.5): *YMS-6*, *YMS-8*, *YMS-9*, *YMS-46*, *YMS-50*, *YMS-52*, *YMS-68*, *YMS-71*, *YMS-340*, *YMS-365*, *YMS-481*
- Service Unit (Task Unit 78.1.6): Sloop HMAS *Warrego* (U73), net laying ship *Cinnamon* (AN-50), fleet tug *Quapaw* (ATF-110), rescue tug *ATR-61*[6]

PRE-ASSAULT LANDING OPERATIONS

The Fire Support (Cruiser Covering Group), Minesweeping, and Hydrographic survey ships, plus motor torpedo boat tender *Oyster Bay* departed Mindoro on 6 March and arrived in the objective area in early morning on 8 March. Pre-assault minesweeping and hydrographic surveys, supported by naval gunfire, were conducted for two days prior to the landings in spite of enemy artillery and machine gun opposition. On 9 March, eight motor torpedo boats arrived, joining *Oyster Bay*, and commenced operations off Basilan Island.[7]

Photo 14-4

Yard minesweeper USS *YMS-71*, location and date unknown, circa 1942. Commanded by Lt. Ernest O. Saltmarsh, USNR, she was the flagship for sweeping operations incidental to the landings at Zamboanga and on Basilan Island.
National Archives photograph #80-G-49330

The YMSs swept the approaches to the Basilan Strait, including close inshore along the south Zamboanga shore. HMAS *Warrego* and USS *Cinnamon* surveyed and planted buoys on Santa Cruz Bank and explored the landing beach area at San Mateo Point. Some counterbattery fire by the Covering Group was required on the 9th, when the minesweeping and hydrographic ships working the north channel received shore fire.[8]

The following entries for 8 and 9 March, related to this action, are from the cruiser USS *Boise*'s report of operations:

8 March 1945

1545: HMAS *WARREGO* firing at troops ashore.

1630: Covering HMAS *WARREGO* and *CINNAMON* (AN 50) planting buoys on Santa Cruz Bank.

9 March 1945

0923: Observed enemy gun fire bearing 053.5°T. Shells landing near *CINNAMON* (AN 50) and HMAS *WARREGO* near Santa Cruz Bank.

0924: Commenced counter-battery fire with main and secondary battery at enemy gun emplacement (0735A). *PHOENIX* opened fire on same target.

0925: Minesweepers in area Half and Half receiving small calibre fire from area 0038M, shifting secondary battery to that area.

0928: Opened fire with secondary battery on area 0038M bearing 013° T. Minesweeper observed returning enemy fire.

0953: HMAS *WARREGO* and *CINNAMON* (AN 50) retired to westward as directed by CTG 74.3 [commander, Cruiser Division Fifteen]. Resumed firing main battery to port at target #42 (0536W).

1435: Sweepers receiving fire from beach to eastward of Caldera Point. Taking station to support destroyers, minesweepers.

1719: HMAS *WARREGO* firing. Her boats conducting survey of landing area.[9]

Photo 14-5

A view from the bridge of HMAS *Warrego* looking aft ship toward the stern. One of her two 28-foot survey motor boats can be seen in its davits at right. *Warrego* did not have an aft mast, so her ensign flies from a jackstaff atop the roof of the cook house. Australian War Memorial photograph PO2305.054

LANDINGS AT ZAMBOANGA

On the morning of 10 August, just prior to the landings by the 41st Infantry Division, minesweepers conducted a final close pass to the beach, protected by their own gunfire and by the scheduled shore bombardment which commenced at 0645.[10]

The first wave of tracked landing vehicles (LVTs)—better known as "Alligators"—beached at 0915 as scheduled. Moderate enemy machine gun, artillery, and mortar fire, which had not been entirely suppressed by naval gunfire and air strikes, commenced about one-half hour later, but was generally ineffective. The assault LVTs were able to mount the steep bank of loose rubble near the water's edge, then moved off toward the flanks to clear room for succeeding waves.[11]

Waves of LSMs (medium landing ships) and LCIs (infantry landing craft) brought in bulldozers, which cut openings in the bank to allow vehicles to pass. After the LSMs and LCIs had cleared the beach, LSTs (tank landing ships) beached in landing slots previously selected. Numerous off-shore coral heads and obstructions limited the number of suitable slots that could be used. In spite of difficulties, about 6,200 troops and one company of medium tanks were landed within the first hour, and by midnight nearly all the assault troops with about 14,000 tons of vehicles and cargo were ashore.[12]

Photo 14-6

Troops disembark from LCIs on Red Beach at Zamboanga, 10 March 1945. Commander Task Group 78.1 (Commander Amphibious Group Six), Amphibious Attack on Zamboanga, Mindanao – Report of, 26 March 1945

Photo 14-7

Tank landing ships on landing beaches at Zamboanga, 10 March 1945. Commanding Officer, USS *Boise*, Report of Operations in Support of Amphibious Landings at Zamboanga, Mindanao, Philippine Islands, 8-18 March 1945, inclusive, 18 March 1945

Photo 14-8

Front of concrete Japanese pillbox (defense fortification) near the beach. Commander Task Group 78.1 (Commander Amphibious Group Six), Amphibious Attack on Zamboanga, Mindanao – Report of, 26 March 1945

During the assault, LSTs *591* and *626*, and LCIs *710* and *779* were damaged by enemy artillery or mortar fire, and an Army LCM was destroyed by a floating mine. Assault troops, aided by naval gunfire for eight days after the landing, and by land-based air support, made good progress at first despite the roads to and within Zamboanga being heavily mined and booby-trapped, and flanked by strong defenses. Resistance increased as troops progressed northward, and continued at enemy positions north of the beachhead as the month ended.[13]

On 16 March, a reinforced company of infantry made a landing on Basilan Island to the south of Zamboanga. It had been estimated that 500 enemy troops were on the island, but the landing was unopposed. Evidence was found that in the past Basilan Island had been used by small Japanese submarines.[14]

UNIT AWARDS

All eleven YMSs comprising the Minesweeping Unit, and the three USN members of the Service Unit at Zamboanga met the criteria for a battle star to adorn the ships' Asiatic-Pacific campaign ribbon on their ribbon boards, and those of crewmembers on their uniform blouses. However, the ships identified by asterisks in the table were ineligible for such award, having previously earned a star for another operation associated with the consolidation of the Southern Philippines. U.S. Navy policy was that military units could receive only a single battle star for any one operation, no matter how many qualifying actions.

The battle star earned by the net laying ship USS *Cinnamon* (AN-50) was her only one during the war.

Consolidation of Southern Philippines: Mindanao Island Landings

Ship	Period	Ship	Period
YMS-6	8-11 Mar 45	*YMS-71**	8-31 Mar 45
*YMS-8**	8-31 Mar 45	*YMS-340**	8-11 Mar 45
YMS-9	8 Mar-4 May 45	*YMS-365*	8-11 Mar 45
YMS-46	8 Mar-23 Apr 45	*YMS-481**	8-11 Mar 45
YMS-50	8 Mar-16 May 45	*Cinnamon* (AN-50)	10-11 Mar 45
YMS-52	10 Mar-4 May 45	*Quapaw* (ATF-110)*	10 Mar-23 Apr 45
YMS-68	8-11 Mar 45	*ATR-61*	10 Mar-23 Apr 45

15

Assault and Occupation of Okinawa

A compilation of preliminary reports of casualties indicates the following for vessels of this Task Group [Mine Flotilla]: KIA 213; WIA 583; MIA 177; TOTAL 973. The total Naval casualties as of 20 June, were KIA 1,105; WIA 3,637; MIA 1,553; TOTAL 6,295. Minecraft therefore suffered 15.45% of the total casualties.

—Rear Adm. Alexander Sharp Jr., USN, commander, Minecraft, U.S. Pacific Fleet, identifying numbers of mine flotilla members killed, wounded, or missing in action as a result of the assault and occupation of Okinawa.[1]

Map 15-1

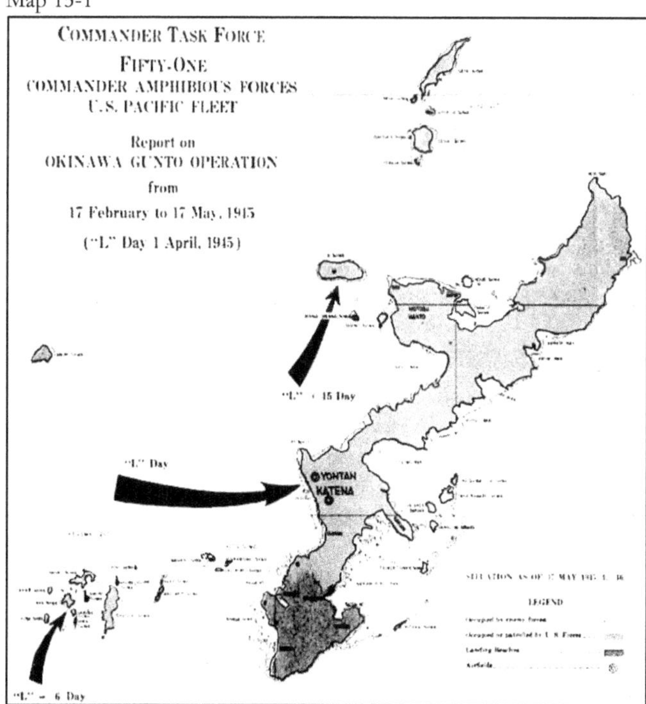

Allied amphibious landings at Kerama Retto on L-6 (26 March 1945), Okinawa on L-Day (1 April 1945), and Ie Shima on L+15 (16 April)

On the morning of 1 April 1945 (L-Day), the U.S. Tenth Army landed over the Hagushi beaches on Okinawa with two corps abreast: the XXIV Army Corps on the southern flank, and the III Amphibious (Marine) Corps on the northern flank. The Northern Attack Force (TF 53) with the 1st and 6th Marine Divisions and the Southern Attack Force (TF 55) with the 7th and 96th Infantry Divisions, landed the assault waves simultaneously, on schedule.

Northern Attack Force (Task Force 53):
Rear Adm. Lawrence F. Reifsnider, USN
III Amphibious Corps (Reinforced): Maj. Gen. Roy S. Geiger, USMC
1st Marine Division, 6th Marine Division

Southern Attack Force (Task Force 55):
Rear Adm. John L. Hall Jr., USN
XXIV Army Corps: Maj. Gen. John R. Hodge, USN
7th Infantry Division, 96th Infantry Division[2]

Okinawa is the largest island of the Okinawa Gunto ("Gunto" means islands), which is part of the Ryukyu Islands. The Ryukyus are a component of the even larger, arc-shaped Nansei Shoto which stretches 790 miles between Kyushu, Japan, and Formosa (today Taiwan). Separating the Pacific from the East China Sea, the Nansei Shoto comprises eight large islands, about twenty smaller ones, and numerous islets, exposed reefs, and rocks. The chain is divided into five groups. Beginning from the south and progressing northward, these are the Sakishima Gunto, Okinawa Gunto, Amami Gunto, Takara Gunto, and Osumu Gunto. The two southern groups are collectively, the Ryukyu Islands, and the three northern ones the Sakishima Islands.[3]

The members of the Okinawa Gunto include Okinawa Shima (island) and many smaller islands, including the Kerama Retto (island group). The action in this chapter begins in this cluster of small islands fifteen miles west of Okinawa. A preliminary operation to capture this roadstead was necessary to gain an advanced naval base for refueling, repairs, and ammunition replenishment necessary to support the largest amphibious force of World War II.

MOVEMENT FORWARD

The Amphibious Support Force—heavy fire support ships, carriers and aircraft, minesweeping vessels, and one section of the underwater demolition group—began departing Ulithi in echelons on 19 March. Mine Group One, the first echelon of Task Force 52 (the Mine Flotilla) arrived at Okinawa Gunto on 24 March, and began minesweeping under the protection of battleship bombardment by Task Force 59 and air cover provided by Task Force 58. The remaining echelons of TF 52 arrived at the objective on 25 March; and the Western Islands Attack Group (Task Group 51.1) arrived at Kerama Retto the following day.[4]

MINESWEEPING REQUIREMENTS AT OKINAWA

Minesweeping preparatory to the assault of Okinawa consisted of the following general components:

- Sweeping necessary for maneuver by pre-assault bombardment units
- Clearance for occupation of Kerama Retto by the Western Islands Attack Group
- Development of swept water to the shoreline for the attack force landings and close fire support
- Resweeping, for maintenance of safe water, of approach routes of Southern Tractor Flotilla and of Transport Groups[5]

To meet these requirements, Rear Adm. Alexander Sharp Jr., commander, Minecraft, U.S. Pacific Fleet (CTG 52.2), assembled an armada of mine warfare ships under his command. The numbers of minesweepers and minelayers (used as mine destruction vessels) that earned battle stars at Okinawa are identified below.

Photo 15-1

Mine Warfare Ships	**No.**
AM Minesweepers	71
YMS Minesweepers	56
DMS Minesweepers	12
DM Minelayers	14
CM Minelayers	3
Total	156

Rear Adm. Alexander Sharp Jr., USN, commander, Minecraft Pacific, 21 January 1945. He was awarded the Distinguished Service Medal on that date in recognition of his service as commander, Battleships Atlantic Fleet and commander, Service Force Atlantic Fleet, during the period of November 1941 to October 1944.
Naval History and Heritage Command photograph #NH 92245

MINESWEEPING AREAS AT OKINAWA, AND THE ASSOCIATED NUMBERS OF MINES DESTROYED

Diagram 15-2

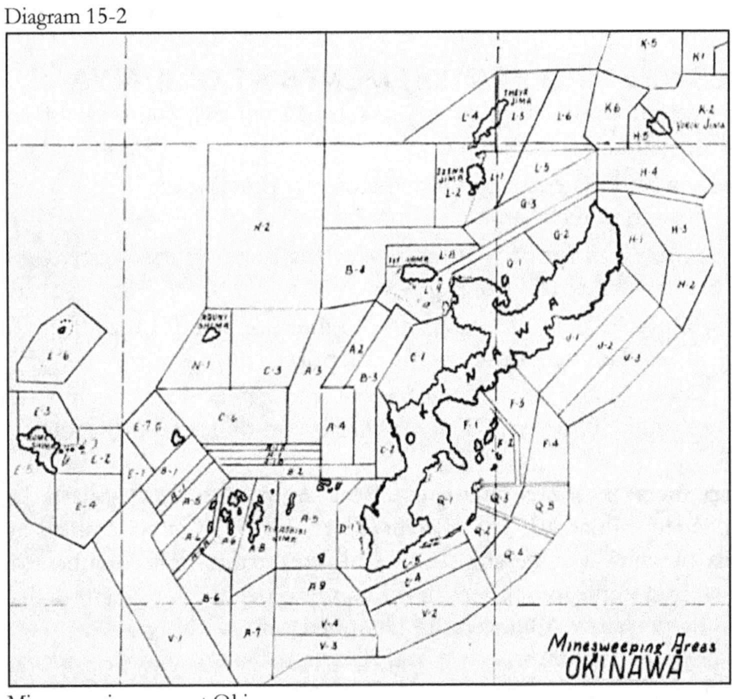

Minesweeping areas at Okinawa
Commander Task Group 32.2 (52.2), Report of Capture of Okinawa Gunto – Phases One and Two, 23 July 1945

A total of 331 (257 + 74) mines were destroyed off Okinawa between 24 March and 9 April 1945. As shown in the table, of this number, 227 (177 + 50) were swept and immediately destroyed. The remaining 104 mines (80 + 24) were "floaters," ones that had broken their moors, or were otherwise freed, found and destroyed at different times.

Swept Mines

L-7 to L-1 (24-30 March 1945)		**L+4 to L+8 (5-9 April 1945)**	
Area(s)	**No.**	**Area**	**No.**
B3 (south), C2	67	Q1	50
B5, A9	26		
C4, C5	36		
William	1		
D2	47		
Total swept mines	177	Total swept mines	50

Floating Mines Destroyed

L-7 to L-1 (24-30 March 1945)		L+4 to L+8 (5-9 April 1945)	
Floaters – various areas	80	Floaters – various areas	24
Total combined swept mines and floaters	**257**	**Total combined swept mines and floaters**	**74**

Although some of the alpha-numeric identifications used on the diagram to identify sweep areas are hard to distinguish, they do correlate with information in the preceding table. Generally, the mines swept before L-Day were found in the southern approaches to Okinawa and off the planned assault beaches on the island's western coast. Those swept after L-Day were in the bay on the southeast coast.

SHIPS SUNK OR DAMAGED AT OKINAWA

The U.S. Navy suffered horrendous losses of ships and personnel at Okinawa, mostly as a result of attacks by Japanese suicide planes. Information in a commander in chief, Pacific Fleet (CincPacFlt) report included in Appendix C identifies 22 ships sunk and 212 (one British) damaged during the period 26 March to 30 April 1945.

Comparison of this report with a later more detailed one by commander, Minecraft, U.S. Pacific Fleet, addressing mine warfare ships, reveals some errors in the above earlier report. A few of the 212 ships identified as damaged were double-counted; they were attacked, and suffered damage, on more than one occasion, as reflected by more than one entry. Also *Champion* (AM-314), *Gladiator* (AM-319), and *Ransom* (AM-283) were not listed, and some entries are conflated. For example, attacks against *Adams* (DM-27) on three different occasions, were combined into a single entry for 24 April. These attacks are separated in the following table. Finally, some of the dates of attacks cited were off by a day; which can easily happen when it is realized that the ships submitting the reports and staff at CincPacFlt Headquarters in Pearl Harbor compiling associated data are on different sides of the International Dateline.

Four of the twenty-two ships sunk and 29 of those damaged were mine warfare vessels, identified in the table. (*Adams* was damaged on three separate occasions, and *Gladiator*, *Harding*, and *Skirmish* twice.) Almost all damage suffered resulted from enemy action. Exceptions were *YMS-96*, involved in a collision with *Hambleton* (DMS-20), and *Tolman* (DM-28) which grounded on a shoal while on an escort mission.

Mine Warfare Ships Lost or Damaged at Okinawa
(26 March-30 April 1945)

Date	Ship	Cause	Extent
23 Mar 45	*Adams* (DM-27) KIA: 2, WIA: 9, MIA: 0	air attack; accident premature burst from 5" round fired by ship	minor
26 Mar 45	*Dorsey* (DMS-1) KIA: 0, WIA: 2, MIA: 3	suicide plane	minor
26 Mar 45	*Robert H. Smith* (DM-23)	air attack	minor
26 Mar 45	*Skirmish* (AM-303) KIA: 0, WIA: 1, MIA: 0	suicide plane	minor
27 Mar 45	*Adams* (DM-27)	suicide near miss	minor
28 Mar 45	*Skylark* (AM-63) KIA: 5, WIA: 25, MIA: 0	mine strike	SUNK
31 Mar 45	*Adams* (DM-27)	suicide plane	major
1 Apr 45	*Skirmish* (AM-303) KIA: 1, WIA: 0, MIA: 0	air attack	minor
3 Apr 45	*Hambleton* (DMS-20)	suicide near miss	minor
5 Apr 45	*Harry F. Bauer* (DM-26)	dud torpedo	minor
6 Apr 45	*Defense* (AM-317) KIA: 0, WIA: 6, MIA: 0	suicide plane	minor
6 Apr 45	*Devastator* (AM-318)	suicide plane	minor
6 Apr 45	*Harding* (DMS-28)	near bomb miss	minor
6 Apr 45	*Recruit* (AM-285)	suicide plane	minor
6 Apr 45	*Rodman* (DMS-21) KIA: 10, WIA: 20, MIA: 6	suicide plane	major
6 Apr 45	*Emmons* (DMS-22) KIA: 17, WIA: 71, MIA: 40	suicide plane	SUNK
6 Apr 45	*YMS-311* KIA: 1, WIA: 2, MIA: 0	suicide plane	minor
6 Apr 45	*YMS-321*	air attack	minor
7 Apr 45	*Ransom* (AM-283)	suicide plane	minor
7 Apr 45	*YMS-427* KIA: 3, WIA: 1, MIA: 0	hit from shore battery	minor
8 Apr 45	*YMS-92* KIA: 0, WIA: 2, MIA: 0	mine	major
8 Apr 45	*YMS-103* KIA: 5, WIA: 7, MIA: 0	mine	SUNK
10 Apr 45	*YMS-96*	operational (collision)	major
12 Apr 45	*Gladiator* (AM-319) KIA: 0, WIA: 1, MIA: 0	suicide near miss	minor
12 Apr 45	*Jeffers* (DMS-27)	suicide plane	major
12 Apr 45	*Lindsey* (DM-32) KIA: 13, WIA: 46, MIA: 43	suicide plane	major
16 Apr 45	*Champion* (AM-314) KIA: 0, WIA: 5, MIA: 0	near bomb miss	minor
16 Apr 45	*Harding* (DMS-28) KIA: 14, WIA: 10, MIA: 8	suicide plane	major
16 Apr 45	*Hobson* (DMS-26) KIA: 4, WIA: 6, MIA: 0	suicide plane	major

18 Apr 45	*Spear* (AM-322)	air attack	minor
19 Apr 45	*Tolman* (DM-28)	operational (grounded)	major
22 Apr 45	*Gladiator* (AM-319) KIA: 1, WIA: 5, MIA: 0	suicide near miss	minor
22 Apr 45	*Shea* (DM-30) KIA: 0, WIA: 2, MIA: 0	air attack	minor
22 Apr 45	*Swallow* (AM-65) KIA: 0, WIA: 9, MIA: 2	suicide plane	SUNK
29 Apr 45	*Butler* (DMS-29) KIA: 0, WIA: 6, MIA: 0	suicide near miss	minor
29 Apr 45	*Shannon* (DM-25)	suicide near miss	minor
30 Apr 45	*Terror* (CM-5) KIA: 42, WIA: 121, MIA: 4	air attack	major
30 Apr 45	*J. William Ditter* (DM-31)	air attack	minor[6]

AIRCRAFT SUICIDE ATTACKS

Map 15-3

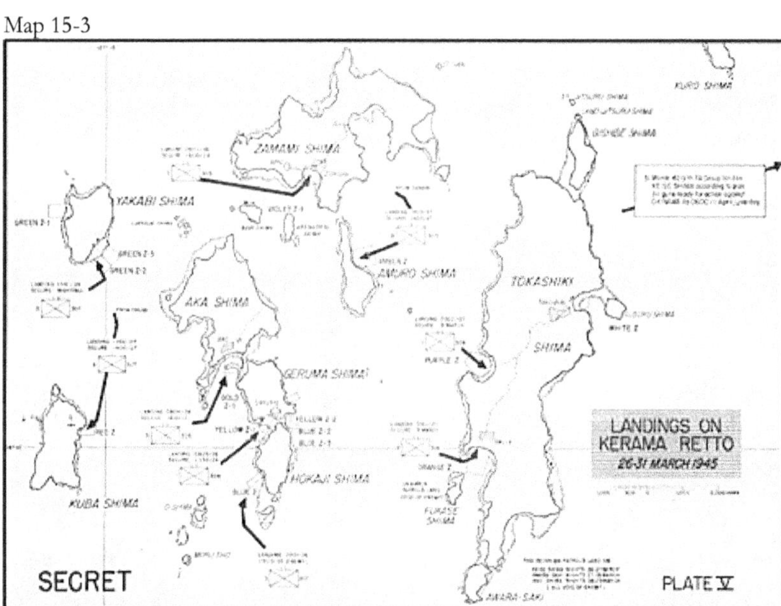

Landings on Kerama Retto Islands by the 77th Division between 26-31 March 1945
Commander in Chief, U.S. Pacific Fleet and Pacific Ocean Areas, Operations in the Pacific Ocean Areas – April 1945, 16 October 1945

Japanese air attacks, by suicide and conventional aircraft, against Allied shipping began immediately following the arrival of minesweepers and minelayers (functioning as mine disposal ships) off Okinawa. The first personnel casualties suffered were aboard the minelayer *Adams* (DM-27). On 23 March, while engaging two enemy aircraft with 5"/38 fire, a premature burst of a round from Mount III caused the deaths of

Seaman 1st Class Harry Franklin Neff and Seaman 2nd Class Charles David Weaver, and wounded nine other crewmembers. The attack was driven off, and at least one of the planes was damaged.[7]

Air attacks continued during landings by the U.S. Army's 77th Infantry Division on islands of Kerama Retto, from 26-31 March 1945, and persisted for weeks during the assault and occupation of Okinawa and Ie Shima, with large Kamikaze raids beginning on 6 April, five days after the landings on Okinawa.

The first loss of a unit of the mine force occurred on the morning of 28 March, when the minesweeper *Skylark* (AM-63) hit a mine and sank in about 30 fathoms of water off Hagushi Beach on the west coast of Okinawa. She had been leading the sweeping formation of Units Six and Seven. The mine blast killed five members of her crew and wounded another twenty-five.[8]

DANGERS ASSOCIATED WITH PICKET STATIONS

Surface forces engaged in and supporting amphibious operations at Okinawa, particularly destroyers assigned to radar picket stations, were frequently hit and some sunk by Kamikazes.

Map 15-4

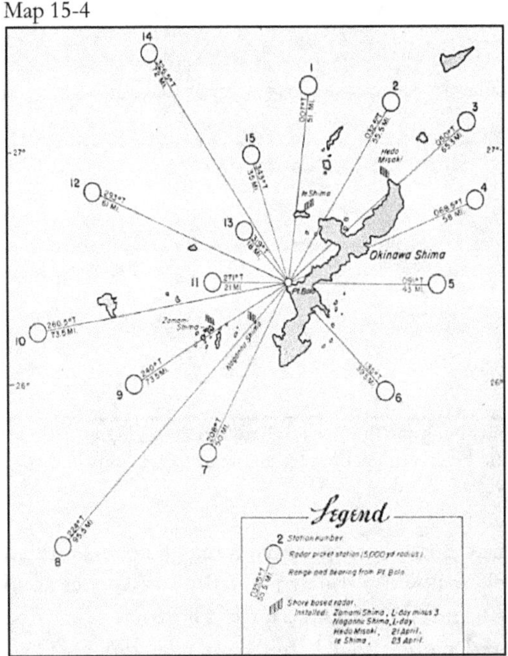

Radar Picket Stations at Okinawa
ComPhibsPac Op Plan A1-45

To provide early detection of enemy aircraft raids arriving from Japanese home islands and bases on Taiwan, the Navy set up a ring of radar picket stations around Okinawa. (Ships were assigned to these stations to detect with their radar and report incoming enemy air raids.) These stations were at distances of eighteen to ninety-five miles from Point Bolo (Zampa Misaki), just north of the Hagushi invasion beaches at the tip of Motobu Peninsula. Ships patrolled various stations from 24 March to 13 August 1945; but stations 6, 8, and 13 were never used. When the Japanese identified the positions of stations, some of them were relocated.[9]

Destroyers, equipped with specialized radar and fighter director teams, were assigned to the stations, as were other vessels to provide supporting fire power. These included additional destroyers (DD), high-speed minesweepers (DMS), light minelayers (DM), medium landing ships fitted with rockets (LSMR), patrol motor gunboats (PGM), and large support landing craft (LCSL).

When a destroyer on picket station detected incoming aircraft, a Combat Air Patrol (CAP), from the carriers and from the fields on Okinawa and later Ie Shima, was notified and sent to intercept the enemy. A flight of CAP was assigned to each picket station. As Japanese airmen discovered that getting through large numbers of CAP aircraft was difficult, they began to select the nearest targets of opportunity which were the ships on the radar picket stations. As a result, the radar picket ships and their escorts bore the brunt of the Kamikaze attacks in what has been described as the most hazardous naval duty of World War II.[10]

Destroyers became obvious targets, and DMS minesweepers and DM minelayers (converted DDs that retained their original silhouettes, and a majority of their guns) were frequently attacked when assigned to picket stations, and when off working with the mine forces.

On 6 April, the worse day for Kamikaze attacks on Mine Force ships at Okinawa, *Emmons* (DMS-22) was sunk and *Rodman* (DMS-21) knocked out of action, both with larger numbers of crewmen killed, wounded in action, or missing in action than in previous attacks. Four other minesweepers suffered minor damage, two of them—*Defense* (AM-317) and *YMS-311*—with some casualties as well. The remarkable story of the survival of *YMS-311* during an attack by multiple suicide planes follows.

YMS-311 STRUCK BY KAMIKAZE, SHOOTS DOWN THREE OTHER ENEMY PLANES IN SAME ATTACK

During a period of two (2) hours, twenty six (26) suicide dives by the enemy were within visual range. Four F4-U's [Corsair fighter aircraft] were seen providing air coverage in the area.

—Lt. Glenn E. Smith, USNR, commanding officer, USS *YMS-311*, in an action report describing combat with Japanese aircraft.[11]

The excellent performance of this vessel is noted with considerable pride. The shooting down of 3 enemy airplanes plus one suicide plane crash, accounted for by this ship, is considered to be a record for this type.

—Rear Adm. Alexander Sharp Jr., USN, commander, Minecraft, U.S. Pacific Fleet.[12]

Photo 15-2

Painting by Richard DeRosset of USS *YMS-311* shooting down four attacking Japanese Navy bombers while engaged in minesweeping off Okinawa.

In late afternoon on 6 April, *YMS-311* was conducting minesweeping operations off the western coast of Okinawa, five miles south of Ie Shima, with port gear streamed. She was steering a westerly course. Six other YMSs and one AM were in the immediate vicinity. Suddenly, without warning, a large group of thirty-five "Vals" appeared from the north, the closest, less than five miles away. The weather was generally cloudy, averaging about one-half of the sky covered, and these conditions helped enemy aircraft to remain hidden in, and to dive from, nearby cloud formations. The wind and seas were relatively calm, with winds of 4-10 knots from the northeast, and ripples on the surface with 1/2-foot wave heights.[13]

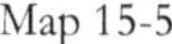
Map 15-5

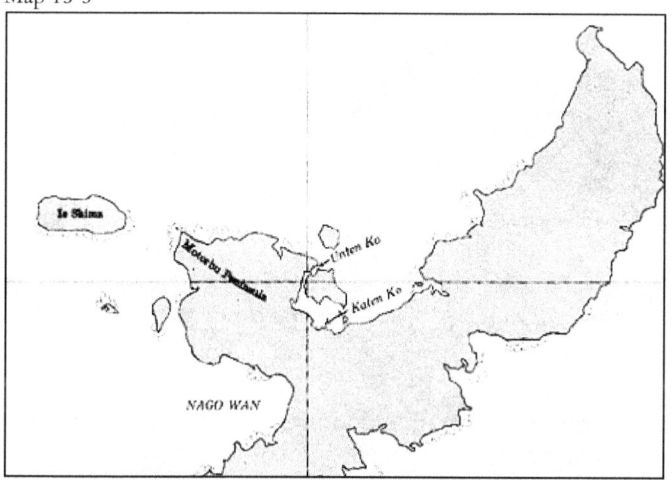

The island of Ie Shima, about three-and-one-half miles off the western tip of Motobu Peninsula, on the northwest coast of Okinawa
https://www.history.navy.mil/content/history/nhhc/research/library/online-reading-room/title-list-alphabetically/b/beans-bullets-black-oil.html

Four of the enemy planes—Aichi D3A Type 99 Navy carrier bombers—chose the *YMS-311* as their target. Coming in from off her starboard side, they sped toward the small, wooden ship at 200 knots, 100 feet off the water. No Flash radio warning had been received prior to the appearance of the aircraft and subsequent attack.[14]

As the General Quarters alarm sounded aboard the minesweeper, and her crew manned their battle stations, the first Val strafed the *YMS-311* with 7.7mm gunfire, then crashed into her foc's'le deck in a suicide attack. No personnel casualties resulted because the deck was clear of crewmen where the strafing was the worst. Damage to the ship by the Kamikaze was minimal. Gunfire from the ship's starboard 20mm and .50-caliber mounts had found the plane, and degraded the pilot's control of it, otherwise it would likely have struck the deckhouse.[15]

The minesweeper was armed with one 3"/50 gun on the foc's'le, two single 20mm anti-aircraft guns on the port and starboard sides of the pilot house, and .50-caliber machine guns on the port and starboard sides of the fantail.[16]

Photo 15-3

Japanese Navy Type 99 carrier bomber (Val) in action during an attack.
National Archives photograph #80-G-32908

The plane hit the foc's'le deck to the right of centerline, rendering the 3"/50 gun mount inoperative, and causing other minor damage:

- 3"/50 breech block carried away, and trunnions sprung out of shape
- Door of port ready ammunition box carried away and trays in box bent; ammunition in the box did not explode
- Railing on both sides of foc's'le and port life raft carried away
- Foc's'le frames sprung slightly from their normal position[17]

Fortunately, there was no resultant explosion or fire, and the wreckage slid into the sea. Damage control consisted of washing down the foc's'le with a fire hose to clear away gasoline and debris.[18]

As three other Vals came in on attack runs, they were all shot down by fire from the ship's 20mm and .50-caliber guns. Two of the aircraft never reached the minesweeper. One crashed into the sea slightly ahead of, and fine on (just off) her starboard bow; a second plane hit the water slightly ahead of her starboard beam, farther from the ship. These aircraft were "splashed" by her starboard gun crews. The third aircraft crossed over her fantail and, as it reversed course, was taken under fire by the port .50-caliber gun. Damaged, it crashed on Ie Shima. The ship's guns were manually elevated and trained, with tracer rounds being the only method of fire control (directing rounds onto the targets).[19]

During these attacks, one crewman—Seaman 2nd Class Hansford W. Connor, USNR—was killed and two others wounded. Within twenty minutes, three destroyers arrived for protection. Five more enemy planes attacked the group of destroyers and minesweepers; all were shot down, but one crashed into a destroyer.[20]

YMS-311 returned to Kerama Retto and transferred her deceased crewmember to a designated ship. On 9-10 April, she swept the Kerama Retto anchorage for moored mines suspected of being laid by enemy aircraft. No mines were discovered. On the evening of 9 April, she patrolled the north entrance of the main anchorage at Kerama Retto.[21]

On 11 April, the minesweeper returned to Saipan and received battle damage repairs from the *Phaon* (ARB-3), which included the replacement of her damaged 3-inch gun.[22]

USS *YMS-311* was later awarded the Navy Unit Commendation, for the heroic actions of her crew on 6 April 1945 at Okinawa.

CAPTURE OF IE SHIMA

On the morning of 16 April, the 77th Infantry Division landed on Ie Shima to secure the airfield on the island. One Regimental Combat Team (RCT) landed over Green Beaches, and one over Red Beaches, while the third RCT was held in reserve. Initially, resistance on the beaches and most of the island was light, but the area surrounding the airfield and the field itself was heavily mined with all types of mines and 300-lb bombs. Enemy resistance increased over the next four days.[23]

On 20 April, the 77th Infantry Division (including the reserve RCT landed on the 17th) began an all-out attack. Organized resistance was overcome on 22 April. This was the same day the Navy lost its fourth, and last Mine Flotilla ship at Okinawa to mines or Kamikaze aircraft.[24]

SWALLOW SUNK BY KAMIKAZE ATTACK

On the evening of 22 April, USS *Swallow* (AM-65) was sunk by a Kamikaze in a single aircraft attack on her, independent of larger raids associated with Operation KIKUSUI ("floating chrysanthemums"). The Kamikaze and conventional aircraft that participated in the initial and largest raid of KIKUSUI, on 6-7 April 1945, consisted of:

- 230 Imperial Japanese Naval Air Force
- 125 Imperial Japanese Army Air Force Kamikazes
- 340 conventional attackers and escorts[25]

Nine more Kikusui raids followed, none as large as the first. Three of these occurred later in April (12-13, 15-16, 27-28); four in May (3-4, 10-11, 24-25, 27-28), and two in June (3-7, 21-22). Altogether, the Kikusui expended about 1,500 Kamikaze planes, and about 400 more were used in small group attacks.[26]

SUMMATION

Enemy air attacks continued during the period 1 May to 21 June. From 1 to 31 May, 728 enemy planes were destroyed in the objective area by TF 51(31) and by TF 58(38). Of the 728 planes destroyed, 302 were shot down by ships' gunfire, 349 by CAP, and there were 77 suicide crashes. During the same period, 9 ships were sunk by enemy air attacks, and 76 were damaged by all types of enemy air attacks. These figures may be compared with 1,124 enemy planes destroyed in the objective area for the period 1 to 30 April, and with 15 U.S. ships sunk and 116 damaged by all types of enemy air attacks.

—Fleet Adm. Chester Nimitz, commander in chief, U.S. Pacific Fleet and Pacific Ocean Areas, citing in his Operations in Pacific Ocean Areas – May 1945 report, numbers of Japanese planes shot down at Okinawa by Task Force 51 (Joint Expeditionary Force) and Task Force 58 (Fast Carrier Force).

The eighty-two-day-long Battle of Okinawa was the largest amphibious assault in the Pacific, one lasting from 1 April through mid-June 1945, and resulting in the highest number of casualties in the Pacific Theater during World War II. The Tenth Army suffered 7,613 killed or missing in action, and 31,800 wounded, while Marine Corps casualties overall—ground, air, and ships' detachments—exceeded 19,500. Navy losses were 34 vessels and craft sunk and 368 damaged, over 4,900 sailors killed or missing in action, and over 4,800 wounded. Japan lost over 100,000 soldiers, either killed, captured or committed suicide.[27]

"KISSING COUSINS" IN ABUNDANCE

As shown in the following section, 56 of the 139 minesweepers that earned battle stars at Okinawa were wooden-hulled yard minesweepers (YMS). Sixteen of their cousins were also awarded stars; 12 patrol craft sweepers (PCS) and 4 hydrographic survey ships (AGS). The latter vessels were former PCSs, reclassified as AGSs and given names. The next chapter is devoted to the contributions of the PCSs and AGSs at Okinawa.

The YMSs that garnered battle stars at Okinawa are identified in the following tables, along with their commanding officers. The Sweep Unit to which individual ships were assigned is also provided for YMSs involved in the initial phases of the assault and occupation. Seventeen YMSs that participated in June only are not associated with any particular Sweep Group or Task Unit.

YMS ORGANIZATION / BATTLE STARS AWARDED

YMS Group (Task Group 52.6): Lt. Comdr. C. A. Bowes, USNR

Sweep Unit Twelve (TU 52.6.1): Lt. Peter F. Beville, USNR

Ship	Award Period	Commanding Officer
YMS-81 (flagship)	25 Mar-30 Jun 45	Lt. Peter F. Beville, USNR
YMS-140	25 Mar-30 Jun 45	Lt. L. A. Agar, USNR
YMS-293	25 Mar-20 May 45	Lt.(jg) A. J. Moore, USNR
YMS-327	25 Mar-30 Jun 45	Lt. George A. Ashland Jr., USNR
YMS-331	25 Mar-9 Apr 45	Lt. John M. Talia, USNR
YMS-341	25 Mar-30 Jun 45	Lt. H. O. Arend, USNR
YMS-410	25 Mar-30 Jun 45	Lt.(jg) Leslie G. Knoles, USNR
Robert H. Smith (DM-23)	25 Mar-30 Jun 45	Comdr. Henry Farrow, USN

Sweep Unit Thirteen (TU 52.6.2): Lt. Louis H. Countryman, USNR

Ship	Award Period	Commanding Officer
YMS-176 (flagship)	25 Mar-30 Jun 45	Lt. Louis H. Countryman, USNR
YMS-219	25 Mar-21 Jun 45	Lt. P. E. Fertig, USNR
YMS-243	25 Mar-28 May 45	Lt.(jg) W. R. Spencer, USNR
YMS-286	25 Mar-30 Jun 45	Lt. D. L. Middleton, USNR
YMS-311	25 Mar-30 Jun 45	Lt. Glenn E. Smith, USNR
YMS-319	25 Mar-30 Jun 45	Lt. Henry G. Bugbee, USNR
Shannon (DM-25)	25 Mar-30 Jun 45	Comdr. R. L. Foster Capt. W. G. Beecher, USN (CoMinDiv 7) embarked

Sweep Unit Fourteen (TU 52.6.3): Lt. H. B. Silveira, USNR

Ship	Award Period	Commanding Officer
YMS-324	25 Mar-30 Jun 45	Lt. H. B. Silveira, USNR
YMS-325	25 Mar-30 Jun 45	Lt. J. M. Ferguson, USNR
YMS-342	25 Mar-28 May 45	Lt.(jg) N. F. Murphy, USNR
YMS-360	25 Mar-30 Jun 45	Lt. R. F. Donovan, USNR
YMS-398	27 Mar-30 Jun 45	Lt. William A. Latta, USNR
YMS-408	25 Mar-30 Jun 45	Lt. Claude Griffin Jr., USNR
PGM-17	25 Mar-30 Jun 45	Lt.(jg) C. W. Cassidy, USNR
Thomas E. Fraser (DM-24)	25 Mar-30 Jun 45	Comdr. R. J. Woodman, USN

Sweep Unit Fifteen (TU 52.6.4): Lt. Comdr. T. E. Lavender

Ship	Award Period	Commanding Officer
YMS-90 (group flagship)	25 Mar-30 Jun 45	Lt. A. G. Dohrmann
YMS-91	25 Mar-30 Jun 45	Lt.(jg) William A. Hurst, USNR
YMS-388	25 Mar-30 Jun 45	Lt. W. S. Ayers, USNR
YMS-423 (flagship)	25 Mar-30 Jun 45	Lt. William D. McAfee
YMS-431	25 Mar-30 Jun 45	Lt. H. J. Bateman
YMS-434	25 Mar-30 Jun 45	Lt. Donald Birdwell, USNR
PGM-18	25 Mar-8 Apr 45	Lt. Cyril Bayly, USNR
Harry F. Bauer (DM-26)	25 Mar-11 Jun 45	Comdr. Richard C. Williams Jr., USN

Reserve Group (TG 52.7): Comdr. E. D. McEathron (CoMinRon 10)
YMS Reserve Group (TU 52.7.1): Lt. R. E. Crowley

Sweep Unit Seventeen (TU 52.7.2): Lt. Charles C. Montgomery, USNR

Ship	Award Period	Commanding Officer
YMS-92 (flagship)	2 Apr-28 May 45	Lt. Charles C. Montgomery, USNR
YMS-93	2 Apr-30 Jun 45	Lt.(jg) A. H. Bofinger, USNR
YMS-103	2-9 Apr 45	Lt.(jg) L. M. Thorton Jr., USNR
YMS-283	2 Apr-30 Jun 45	Lt. J. W. Holmes, USNR
YMS-299	2 Apr-30 Jun 45	Lt.(jg) O. Sipari, USNR
YMS-427	2 Apr-30 Jun 45	Lt.(jg) G. L. Denton Jr., USNR

Sweep Unit Eighteen (TU 52.7.3): Lt. R. E. Crowley

Ship	Award Period	Commanding Officer
YMS-1	1 Apr-18 Jun 45	Lt. R. N. Compton, USNR
YMS-2	1 Apr-18 Jun 45	Lt. C. P. Rafferty, USNR
YMS-86	2 Apr-21 Jun 45	Lt.(jg) Jack W. Welty, USNR
YMS-89	2 Apr-18 Jun 45	Lt.(jg) W. L. Ford, USNR
YMS-94	2 Apr-18 Jun 45	Lt. J. R. Stanfield, USNR
YMS-96	2 Apr-5 May 45	Lt.(jg) W. R. Raleigh, USNR
YMS-183 (flagship)	2 Apr-30 Jun 45	Lt. Stanley C. Klein, USNR
YMS-271	2 Apr-30 Jun 45	Lt.(jg) C. J. Verplank, USNR

Participation in June Only

Ship	Award Period	Commanding Officer
YMS-99	27-30 Jun 45	Lt.(jg) D. H. Evans, USNR
YMS-177	27-30 Jun 45	Lt.(jg) E. A. Brown, USNR
YMS-276	27-30 Jun 45	Woodward Romaine
YMS-323	2-30 Jun 45	Lt. J. R. Thompson, USNR
YMS-343	27-30 Jun 45	Lt.(jg) Charles W. Hatch, USNR
YMS-362	18-30 Jun 45	Lt. T. N. I. Davis, USNR
YMS-372	29-30 Jun 45	unknown
YMS-376	29-30 Jun 45	Lt. Roger L. Davis, USNR
YMS-390	27-30 Jun 45	Lt.(jg) R. A. Eaton, USNR
YMS-401	27-30 Jun 45	Lt. Arleigh P. Hess Jr., USNR
YMS-415	26-30 Jun 45	Lt. S. H. Moore, USNR
YMS-426	27-30 Jun 45	Lt. M. R. Ball Jr., USNR
YMS-443	27-30 Jun 45	Lt. John P. Hanna, USNR
YMS-461	27-30 Jun 45	Lt.(jg) C. M. Kirkham, USNR
YMS-467	26-30 Jun 45	Lt. C. E. Snyder, USNR
YMS-471	26-30 Jun 45	Lt. R. C. Battey Jr., USNR
YMS-475	26-30 Jun 45	Lt. W. E. Bearden Jr., USNR

16

PCSs and AGSs at Okinawa

On Love Day (1 April 1945) while standing in to take station off our assigned beaches two enemy suicide planes were shot down by AA fire from nearby ships about 1,000 feet overhead and both crashed into the sea close aboard. It appears that a light cruiser was the plane's target.... Prior to How [H] hour a most effective shore bombardment was conducted by own fire support ships and the sea wall which would have presented quite a problem in moving supplies inland, had huge holes blown in it and the problem was all but eliminated. There was no fire received from the beach whatsoever and the 7th Division of the 10th Army received no opposition during the entire first day, according to reports received.

—Commanding Officer, USS *PCS-1452*, Action Report for Invasion of Okinawa Jima, 9 April 1945.

Photo 16-1

4 April 1945, Okinawa. USS *PCS-1452* is in the lower left corner of photograph.
U.S. Army Signal Corps photograph SC 207 486

Okinawa was the crescendo of World War II regarding the numbers of "kissing cousins" that participated in the massive amphibious operation. Seventy-two, 136-foot wooden vessels, born in American shipyards and built on the same hulls, but with some differences based on their unique roles, earned battle stars at Okinawa.

- 56 Yard Minesweepers (YMS)
- 12 Patrol Craft Sweepers (PCS)
- 4 Hydrographic Survey Ships (AGS)

PCS / AGS SUMMARY INFORMATION

> *Plans required the use of 54 Control Vessels, expanded from the originally assigned. 6 PCS's and 6 PC's, which had previously been partially converted, had to be outfitted in the forward area.*
>
> —Commander Amphibious Forces, U.S. Pacific Fleet. Twelve of the fifty-four control vessels required at Okinawa were PCSs.[1]

The patrol craft sweepers (PCSs), and hydrographic survey ships (AGSs which then had been recently converted from PCSs) that earned battle stars for the Invasion and Occupation of Okinawa are identified below, along with their commanding officers.

Patrol Craft Sweepers

Ship	Award Period	Commanding Officer
PCS-1379	26 Mar-3 Jun 45	Lt. E. U. O'Donnell, USNR
PCS-1389	1 Apr-24 Jun 45	Lt. A. L. Goldstein, USNR
PCS-1391	1 Apr-30 Jun 45	Lt. Robert F. Stafford, USNR
PCS-1402	1 Apr-7 Jun 45	Lt R. G. Spencer, USNR, Lt. (jg) J. R. Taylor Jr., USNR
PCS-1403	1 Apr-30 Jun 45	Lt. John M. Cherry Jr., USNR Lt. Robert R. Eckart Jr., USNR
PCS-1418	1 Apr-30 Jun 45	Lt. K. T. Manning, USNR
PCS-1421	1 Apr-22 Jun 45	Lt. Edmond T. Freeman, USNR
PCS-1429	1 Apr-24 Jun 45	Lt. (jg) W. W. Beardsley Jr., USNR
PCS-1452	1 Apr-30 Jun 45	Lt. (jg) J. S. Simms, USNR
PCS-1455	1 Apr-30 Jun 45	Lt. Carter F. Cort, USNR
PCS-1460	1 Apr-27 Jun 45	Lt. Foster William Lamb, USNR
PCS-1461	1-14 Apr 45	Lt. Willis S. Harrison, USNR

Hydrographic Survey Ships (Former Patrol Craft Sweepers)

Littlehales (AGS-7) ex-USS *PCS-1388*	1 Apr-28 Jun 45	Lt. (jg) Edwin J. Richards, USNR
Dutton (AGS-8) ex-USS *PCS-1396*	1 Apr-7 Jun 45	Lt. Frederic E. Sturmer, USNR
Amistead Rust AGS-9) ex-USS *PCS-1404*	26 Mar-30 Jun 45	Lt. (jg) John G. Carlson, USNR
John Bliss (AGS-10) ex-USS *PCS-1457*	26 Mar-30 Jun 45	Lt. Frank A. Woodke, USNR

PREPARATIONS FOR PCS CONTROL VESSELS

Advance preparations, to equip and man vessels for performing amphibious landing control operations for the Okinawa Operation, were completed at Pearl Harbor prior to sailing for Okinawa. Vessels sailing from Pearl Harbor assigned to the Southern, Northern, and Western Assault Groups, and Demonstration Group, joined their Task Force or Task Group at the locations indicated in the table.

Control Group	No. Vessels	CTF/CTG	Location
Southern	20 (mainly from Iwo Jima)	CTF 55	Leyte
Northern	19	CTF 53	Guadalcanal
Western	8	CTG 51.1	Leyte
Demonstration	7	CTG 51.2	Saipan[2]

Specially equipped patrol craft sweepers were assigned to each group, except the Demonstration Group (decoy group), as flagships. Lessons learned by the control group at Iwo Jima were incorporated in the planning, training, and operations of the other groups. Moreover, experienced key men and officers were assigned to each of these groups, as they were formed and before the landings were rehearsed for the Okinawa operation at Pearl Harbor.[3]

Special Control Communication Teams were organized for the PCSs prior to their change-over to control vessel duties. These teams received a special four-week course ashore under experienced officers, before being embarked on board to begin their basic amphibious training. The key requirement for control vessels was to have efficient, well trained, and disciplined combat control communications teams, which were the heart of the control system. As part of this, every effort was made to eliminate unnecessary radio transmissions and call-ups.[4]

As shown below, six of the twelve PCSs were assigned to the Southern Attack Force, five to the Northern Attack Force, and the remaining one to the Western Islands Attack Group.

Southern Attack Force (TF 55): Rear Adm. John L. Hall Jr.

55.4 (Southern Control Group): Capt. Bruce Bryon Adell
55.4.1 (Southern Flagship Unit): Capt. Bruce Bryon Adell
PCS-1402, 1421
55.4.2 (Control Unit Dog): Lt. Comdr. Dundon
PCS-1403, PCS-1452
55.4.5 (Control Unit Easy): Lt. Trappe
PCS-1455, PCS-1461

Northern Attack Force (TF 53): Rear Adm. L. F. Reifsnider

53.4 (Northern Control Group): Capt. Beverly M. Coleman
53.4.1 (Central Control Unit): Capt. Beverly M. Coleman
PCS-1389
53.4.2 (Control Unit Able): Lt. Nisbet
PCS-1460, PCS-1418*
53.4.5 (Control Unit Baker): Lt. Comdr. March
PCS-1391, PCS-1429

Western Islands Attack Group (TG 51.1): Rear Adm. I. N. Kiland

51.1.11 (Western Islands Control Unit): Lt. Baker
Control Unit Three: Lt. E. U. O'Donnell, USNR
PCS-1379[5]

L-DAY (1 APRIL) AND SUBSEQUENT DAYS

As indicated in the quoted material at the chapter's head, the Southern Attack Force, to which USS *PCS-1452* was assigned, encountered almost no enemy opposition during the initial landings on Okinawa on 1 April 1945. However, this lull eventually came to an end, as further reported by the patrol craft sweeper:

> Enemy planes were overhead all the first night and most of the nights of L plus 1 and L plus 2 days. On L plus 3 we had sporadic alerts and on L plus 4 no raids were experienced. On L plus 5 [6 April] the enemy made a frantic attempt to inflict heavy damage on our shipping.
>
> At 1608…an enemy plane (apparently a Val) dove out of the overcast sky overhead, and selected as a target for his suicide dive the *LST 788* which was close aboard this vessel. The plane was set afire by AA fire from the *LST 788* and other nearby vessels and crashed into the water about 10 yards off the starboard quarter of the LST – about 200 yards from us. There was nothing left of the plane on the surface of the water and only an oil slick where he hit.[6]

At 1616, as the action continued, sightings were made aboard the PCS of two planes under AA fire; one identified as enemy and the other as an American carrier-based F6F Hellcat fighter aircraft. Both were shot down in flames, the latter by friendly fire over Okinawa. A short time later, an enemy suicide plane was observed under fire, about two miles to the northwest. It went into a long glide and, in spite of terrific AA barrage, was believed to have hit a large transport. *PCS-1452* fired on this aircraft with her main AA battery.[7]

A little later, another Japanese suicide plane was observed at 1708, in a dive about five miles distant to the northwest. This aircraft was hit by AA fire and crashed into the sea. Subsequent reports coming in conveyed that many DDs, APDs, AMs, and other vessels had been hit by suicide planes (in addition to those incidents witnessed by *PCS-1452*), particularly those on screening stations (radar pickets).[8]

Enemy air attacks continued in the days that followed as evidenced by the experiences of other patrol craft sweepers.

DUTY AS A PRESS SHIP

On the morning of 1 April, responding to an order, *PCS-1403* closed the attack transport *Harris* (APA-2) in the transport area off Hagushi. After Maj. Gen. Archibald V. Arnold, USA, commanding general, 7th Infantry Division, decided not to use her, the patrol craft sweeper received a group of war correspondents on board for a ringside seat of the invasion of Okinawa. They remained on board until the next day, when she procured an LCVP to take the correspondents into the beach.[9]

On 6 April, while anchored near the amphibious force command ship *Estes* (AGC-12), *PCS-1403* experienced her first daytime air attack. Five Japanese planes were shot down within a radius of about 4,000 yards. Only one was close enough and high enough for her to shoot at without endangering nearby shipping. Alas, the plane came through the overcast and crashed alongside a tank landing ship so rapidly that the PCS was unable to bring any guns to bear on it.[10]

In early afternoon on 12 April, the patrol craft sweeper was entering harbor from Nakagusuki Wan when an Oscar was sighted nearly astern, slightly to the left, bearing 220° relative. She executed a 90° turn to port to bring her 3"/50 gun to bear, but by then the plane had passed out of sight. A 1500, a plane of the same type returned, which the ship was able to take under fire. A near miss was scored on the first shot using a 4-second fuze, and several others of the nine rounds expended were close, but no damage to the plane resulted. The plane circled out of range then began coming in from the east. It appeared that the ship was going to get a "working over," but then the pilot apparently changed his mind, and his course, and headed out to sea.[11]

HYDROGRAPHIC SURVEY UNITS

Two of the four hydrographic survey units at Okinawa—*Littlehales* (AGS-7) and *Dutton* (AGS-8)—had been engaged in survey work at Iwo Jima. The other two—*John Blish* (AGS-10) and Stockton-built *Armistead Rust* (AGS-9)—were utilized at Iwo solely for patrol and escort duties. During transit to Okinawa, two of these ships were assigned to Amphibious Group Seven, and two to Group Twelve as escorts. Upon arrival in the operating area, they were released from escort and screen duties and employed solely for survey duties.[12]

Reaching Kerama Retto with the Western Islands Attack Force, the two ships of Lt. Frank Woodke's Hydrographic Survey Unit—*John Blish* (AGS-10) and *Armistead Rust* (AGS-9)—began survey operations in that area. (Woodke was the commanding officer of *John Blish*).

51.17 (Hydrographic Survey Unit): Lt. Frank A. Woodke, USNR
John Blish (AGS-10): ex-*PCS(H)-1457*
Armistead Rust (AGS-9): ex-*PCS(H)-1404*[13]

The survey ships laid additional radar buoys to augment ones previously emplaced by commander, Minecraft, U.S. Pacific Fleet units, checked the positions of those previously placed; observed currents and tides; and examined beaches and anchorages. While conducting hydrographic operations in the Kerama Retto, *Armistead Rust*—the former *PCS-1404/PCS(H)-1404*—and *John Blish* printed and distributed temporary charts showing new developments.[14]

On their departure to take up new duties at Okinawa, all unfinished work and records were transferred to the much larger, steel-hulled *Bowditch* (AGS-4). She later assembled all data developed by units into a field chart of Kerama Retto.[15]

The remaining two survey ships—*Littlehales* (AGS-7) and *Dutton* (AGS-8)—arrived off the Southern Landing Beaches at Okinawa and began hydrographic survey operations on L-Day.[16]

Littlehales (AGS-7): ex-*PCS(H)-1388*
Dutton (AGS-8): ex-*PCS(H)-1396*

They established navigation lights, marked all known offshore dangers, and sounded the waters from Zampa Misaki to Chatan between the low water mark and the 30-fathom contour (180-foot depth water). During this process, several uncharted coral heads were found inside the 10-fathom contour and marked. On completion of soundings in this area, an anchorage chart was prepared, printed, and issued under the authority of commander, Amphibious Forces Pacific.[17]

DUTTON HIT BY KAMIKAZE AIRCRAFT

> *It is urgently recommended that consideration be given to improving AA defense on the eight additional hydrographic vessels of this type now under conversion.*
>
> —Commodore Leon S. Fiske, commander Service Squadron Twelve.[18]

Photo 16-2

USS *Dutton* (AGS-8). The 3"/50 gun forward and the large hull number was typical of these type ships operating in the Western Pacific 1945. Naval History and Heritage Command photograph NH 84644

As the days and weeks passed, *Dutton* operated off bitterly contested Okinawa, conducting surveys by day, and guarding against enemy suicide boats and swimmers at night.[19]

On the morning of 27 May, the survey ship was attacked by three enemy planes while she was en route from Nakagusuku Wan (which American forces called Buckner Bay) on the southeast coast of Okinawa, to Chimu Wan farther up the east coast. She had left anchorage at 0635, and stood out the channel of Buckner Bay. There was slight rain, visibility was reduced, and aboard ship Condition II was set and AA batteries (two 20mm guns) were manned.[20]

Map 16-1

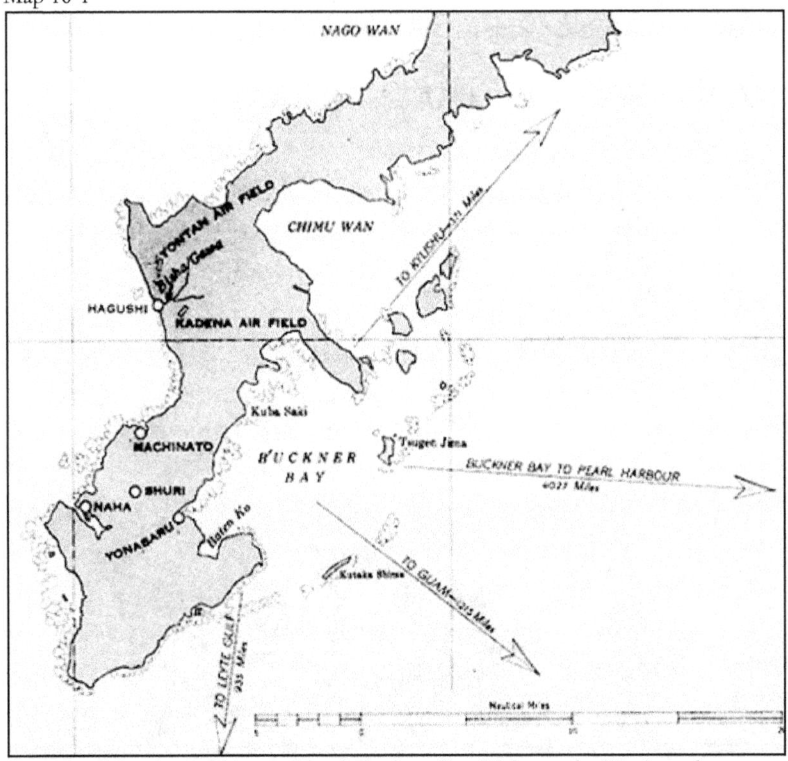

Central and Southern Okinawa Island; Buckner Bay (Nakagusuku Wan) on the southeast coast of Okinawa, and Chimu Wan to the north

At 0735, when three miles off Tsuken Light on Tsuken Island in Nakagusuku Wan, "bogies" (unknown aircraft) were reported to the northeast, ten miles from *Dutton*'s position. Other ships in sight were a DM and a DE about two miles to the north and east, respectively, with several transports and small craft to the south.[21]

At 0740 the lookout reported three planes to the north, just discernable in the clouds at 40 degrees above the horizon. Owing to the poor visibility and fact that the aircraft were flying in a direction away from an American airfield, they were tentatively identified as "friendlies" sent out to intercept the "bogies." Almost immediately they changed to a southerly course and toward *Dutton*, permitting a proper view of their profile, and were immediately identified as Vals.[22]

Within a minute, the planes were over the survey ship and two broke formation and began an attack. The first one crossed over her stern and was taken under fire by the starboard 20mm gun as soon as within range. This plane banked away from the tracer and headed in the direction of the DE; thirty rounds were fired and no hits observed.[23]

Meantime, the second plane commenced a steep dive from just abaft the port beam of the ship. The port 20mm opened fire immediately, and the plane dove straight down the line of fire from 80 degrees above the horizon. No strafing was observed, and it was believed that the ship's fire killed the pilot, for the plane, which was headed for the midships section, swerved off enough to overshoot his mark. The aircraft tore away the ship's flying bridge before striking the starboard side of the foc's'le deck, creating a 4 by 3 by 3-foot hole in the deck, deck edge, and hull planking. Fortunately, momentum took the plane wreckage over the side before the bomb it carried exploded.[24]

Based on the resultant surge of the ship, and deluge of water which covered it entirely with a heavy wash of water, oil, gasoline, and plane parts, the detonation was believed to have occurred at a considerable depth below the ocean's surface.[25]

Following the Kamikaze crash into *Dutton*, her engineers took rapid measures to minimize expected damage as described by Lt. Frederic E. Sturmer, USNR, the ship's commanding officer:

> All hands reacted to the emergency with courage and coolness. The ship was dead in the water. The engineers had stopped both engines believing that the bow had been blown off. The electricians pulled the main switch to prevent short circuits and fires. Within a few minutes the condition of the ship was checked and while there was considerable minor damage, the seaworthiness was not vitally affected. Power was cut in, the engines started and the ship was brought sharply about and headed for the area of the large transports nearby.[26]

Dutton proceeded a mile on this course before a muster of the crew revealed that one man was missing, believed to have been blown or washed overboard by the force of the crash and explosion. Course was reversed and the ship returned to the area of attack, identified by floating debris. A 30-minute search by *Dutton*, and ensuing one by a sub-chaser, *SC-1338*, which arrived on scene, failed to find any trace of the missing crewmember, who was presumed lost.[27]

Dutton returned to the Nakagusuku Wan anchorage, and later underwent temporary repairs at Kerama Retto, before proceeding to San Pedro Bay, Leyte, for permanent repairs.[28]

SOPA KERAMA RETTO

On 2 April 1945, Rear Adm. Ingolf N. Kiland assumed the duties of Senior Officer Present Afloat (SOPA) Kerama Retto (CTG 51.15), having transferred command of Task Group 51.1 to Commodore Thomas B. Brittain, USN. In his new role, Kiland was responsible for the operations of, and Naval defenses for the Fleet Logistics Base, Kerama Retto. These duties had already become well established since the initial landings for capture of the island group on 26 March. The command comprised the islands of the Kerama Retto, enclosing an inner anchorage protected by a series of light anti-torpedo nets, with surrounding anchorage areas used for berthing additional ships which could not be accommodated inside.[29]

Naval vessels were assigned for minesweeping, anti-submarine patrol, and screening of the western anchorage areas, for small boat patrols and general gunboat duties, and for net maintenance. Additionally, LSD (dock landing ship), ARL (landing craft repair ship), ARS (salvage ship), and ATR (rescue tug) types were established in a well-sheltered part of the anchorage, engaged in emergency repairs of ships which suffered battle damage, and ultimately, in routine repairs and maintenance as well.[30]

17

Net Layers, Salvage Ships, and Rescue Tugs at Okinawa

Okinawa Operation: Assault and Occupation of Okinawa
Battle Star Recipients built in Stockton Yards

USS *Snowbell* (AN-52)	26 Mar 45-unknown	Pollock-Stockton Shipbuilding
USS *Spicewood* (AN-53)	26 Mar-30 Jun 45	Pollock-Stockton Shipbuilding
USS *Winterberry* (AN-56)	26 Apr-30 Jun 45	Pollock-Stockton Shipbuilding
USS *Anchor* (ARS-13)	8-30 June 45	Colberg Boat Works
USS *ATR-51*	1-28 Apr 45	Colberg Boat Works
USS *ATR-53*	30 Jun 45	Colberg Boat Works
YH-4 (hospital boat)	24-30 Jun 45	Stephens Bro. Boat Builders

NET LAYING ACTIVITIES

During the morning on two occasions, observed enemy dive bombers make suicide dives on destroyers in the screen. In each case plane came out of the clouds on about 45° glide. The first attacker struck his target in what appeared to be amidships and instantly enveloped the ship in flames. The second, as a result of radical maneuvers by the destroyer, overshot his target and splashed. Each dive bomber was taken under fire before or as he emerged from the clouds, but from our position, without visible results.

—Comdr. George C. King, USNR, commanding officer of USS *Salem* (CM-11), describing in a war diary entry for 26 March 1945, first exposure of the Net and Buoy Group to combat at Okinawa, following its arrival at Kerama Retto, Ryukyu Islands.[1]

In early evening on 19 March 1945, Task Unit 51.7.3 (Net and Buoy Group) took departure from San Pedro Bay, Philippines, bound for Okinawa. The task unit, sailing in convoy as part of Task Group 51.7, consisted of the following eight ships:

- Flagship minelayer USS *Salem* (CM-11)
- Net cargo ship USS *Keokuk* (AKN-4)
- Net laying ship USS *Abele* (AN-58)

- Net laying ship USS *Corkwood* (AN-44)
- Net laying ship USS *Snowbell* (AN-52)
- Net laying ship USS *Spicewood* (AN-53)
- Net laying ship USS *Stagbush* (AN-69)
- Net laying ship USS *Terebinth* (AN-59)[2]

Photo 17-1

Minelayer USS *Salem* (CM-11) under way near Norfolk, Virginia, 29 April 1944. National Archives photo 80-G-229561

The passage to Okinawa was uneventful. In early morning on 26 March, Task Unit 51.7.3 then redesignated Task Group 52.8 was detached from Task Group 51.7, and proceeded to an assigned area off Kerama Retto, Ryukyu Islands, arriving there at 0633. As indicated in the quoted material, the sailors aboard the ships had a ringside view, on their first day in the Okinawan Islands, of combat in progress close at hand. Two of the net laying ships there—*Spicewood* and *Snowbell*—were products of Pollack-Stockton shipbuilding. Although little information exists about *Spicewood*'s activities, that pertaining to *Snowbell* can be found in her ship's history submitted on 29 January 1946, including the leadup to her presence at Okinawa.[3]

PREPARATIONS FOR WAR DUTY

Following her commissioning in Stockton on 16 March 1944, and subsequent shakedown, *Snowbell* was ordered to San Pedro, California, where she was assigned to maintain the extensive harbor net installation there. In addition to these duties, she was utilized by the Small Craft Training Center at San Pedro as a training ship for officers and enlisted personnel assigned to the school.[4]

Photo 17-2

Net laying ship USS *Snowbell* (AN-52).
U.S. Navy photograph

Duty stateside was pleasant for her crew, but net operation requirements in the Pacific soon called her to join sister ships already attached to the Pacific Fleet. By Christmas Eve, *Snowbell* was docked at the Craig Shipbuilding Corporation yard, in nearby Long Beach, for alterations and fitting out in preparation for her mission ahead. With mainmast removed, two 20mm machine guns added, and a new ship's office and sickbay installed, she proceeded up the coast to the San Francisco Bay. After passing under the Golden Gate Bridge, inbound from sea on 22 January 1945, it was only a short leg northward to the Naval Net Depot at Tiburon.[5]

With her holds and foc's'le fully loaded with nets and moorings, *Snowbell* departed Tiburon on 27 January 1945, and stood out of the bay, bound for Pearl Harbor. Once past the Farallon Islands, just off the San Francisco coast, she encountered seas and winds of gale intensity, introducing long rolls and continuous pitching—resulting in the appearance of many buckets at the various watch stations. After arriving at Pearl Harbor on 6 February, the net laying ship remained there for a week, during which time all hands turned to loading stores, taking advantage of this last opportunity to stock up at a well-provisioned rear area base.[6]

Photo 17-3

USS *Snowbell* (AN-52) at Tiburon, California, with nets on deck, which would later be laid in the Invasion of Okinawa.
U.S. Navy photograph

After leaving Hawaiian waters and heading west, *Snowbell*'s next stop was at Johnson Island, Eniwetok Atoll. This was for only a day to take on fuel and water. Continuing onward, her crew was informally initiated on 20 February, into the Order of the Golden Dragon when the ship crossed the 180th Meridian, also known as the International Date Line. With the vast increase in Navy activity in this area in World War II and later, the passage became so common to sailors that few initiation ceremonies were held.[7]

Photo 17-4

DOMAIN of the GOLDEN DRAGON

Ruler of the 180 Meridian

Domain of the Golden Dragon certificate.

On 6 March, *Snowbell* arrived at the then secret base at Ulithi Atoll in the Caroline Islands, where she remained until 11 March when she got under way for Leyte, Philippine Islands. At San Pedro Bay, Leyte, on 15 March, final ordnance and stores were brought aboard. Four days later, *Snowbell* stood out from the bay on the 19th and proceeded north to Okinawa. Love Day (the date of the initial assault landings at Okinawa) was set for Easter Sunday, 1 April.[8]

Her job at Okinawa was to lay the nets she had on board in Kerama Retto, an island group some twenty miles west of that island. In the anchorage where the nets were to be laid, various repair tenders and supply ships were to operate until the larger harbors in Okinawa were secured. On 27 March after a two-day wait offshore, *Snowbell* together with the other net layers in the group, commenced installing a curtain of nets at the south entrance of Kerama Retto necessary to protect the ships there from possible submarine attack.[9]

DETAILS OF NET LAYING

Snowbell made passage from Leyte as part of Task Unit 51.7.3 (Net and Buoy Group) within the larger Convoy Task Group 51.7. After the convoy arrived off Okinawa, Task Unit 51.7.3 (redesignated Task Group 52.8) proceeded to Kerama Retto. A war diary entry by Comdr. George C. King, USNR (commanding officer of USS *Salem* and of the task group), provides a well stated description of the intricacies of laying net on 27 March:

> At 1311 a marker line of buoys was laid. Nets were laid normal to the marker line by dropping a downstream anchor while moving slowly ahead. The downstream mooring payed out until the end was reached then the net started paying out. The upstream mooring was shackled to the end of the net section, so the upstream mooring payed out with the net. Then the anchor end of the upstream mooring took a strain, the anchor stopper was cut, dropping it, as the free end of the net floated clear. Moorings were laid 960 feet apart and when the free end of a section was married to an adjacent section, a continuous net baffle resulted.[10]

Despite almost ever-present raids by Japanese Kamikaze aircraft, this net and others were laid as planned. In April and into May, it was *Snowbell*'s assignment to maintain the net at the north entrance of the anchorage by day (opening and closing it to permit passage by friendly vessels), and serve as a guard vessel at night. Anchoring near the island of Aka Shima, on which enemy were known to be present, created a requirement for several personnel to be topside at all times, roving the deck on the lookout for possible suicide swimmers. These men were in addition to those assigned to man the 20mm machine guns.[11]

SNOWBELL SHOOTS DOWN JAPANESE PLANE

> *During the latter part of May, the attacks by Kamikaze planes became more intense. It was on 25 May that the* SNOWBELL *shot down a Tony – a fast single engine fighter plane – which crashed into the water a few hundred feet from the ship.*
>
> —Ship's History of USS *Snowbell* (AN-52), 29 January 1946.[12]

Photo 17-5

"Tony 1" Kawasaki Ki-61 Hien Army type 3 fighter.
Japanese Operational Aircraft "Know Your Enemy!"
CinCPac - CinCPOA Bulletin 105-45

On 15 May, with the motorless net gate vessel *YNg-30* in tow, *Snowbell* proceeded from Kerama Retto to Okinawa. A net installation was planned for Nakagusuku Wan, later called Buckner Bay by American servicemen, in honor of Army Lt. Gen. Simon Bolivar Buckner Jr., who died on 18 June 1945. He was killed in action while visiting a forward observation post near the southwest tip of Okinawa. *Snowbell* had been relieved of her duties in Kerama Retto by another net layer so that she could take part in the new operation.[13]

During the latter part of May, while anchored close to the shoreline, *Snowbell* was in the foreground during Kamikaze attacks, frequently being the first ship to open fire at oncoming enemy planes. On the 25th, as described in the quoted material, she shot down a "Tony," the Allied code name for a Kawasaki Ki-61 Hien Army type 3 fighter.[14]

WINTERBERRY CLAIMS SIX SURE ASSISTS

Other net laying ships were also in combat with attacking aircraft. *Winterberry*, built by Pollock-Stockton Shipbuilding, a sister ship of *Snowbell*, claimed a collective six "sure assists" related to Japanese planes shot down during raids on 28 May, 3 June, and 11 June. Commanded by Lt. S. E. Aarens, USNR, *Winterberry*'s armament, like *Snowbell*'s, consisted of one 3"/50 gun mount, two 20mm machine guns, and two smaller .50-caliber machine guns. Aarens' report of the action on 28 May includes references to the enemy pilots' tactics. These were attacking from out of the sun or from cloud cover; coming in very low; or some combination of these; and, for multiple-plane attacks, striking from different directions. Planes damaged by anti-aircraft gunfire did not passively dive or fall into the water, but instead sought to crash into the largest, most valuable ships reachable.

> First attack occurred at 0500 with a single float plane crashing a Liberty ship. Condition Blue [caution] existed at the time and no shots were fired by any ships in the harbor. A very successful sneak attack!
>
> At 0730, the first dive bomber appeared very low over the island and crashed into an APA [amphibious attack transport]. Almost all ships fired on this plane although many ships were in line of fire due to this bogey's extreme low approach. This vessel was hit by two (2) 20MM shells from an adjacent merchant AK [cargo ship].
>
> Clouds were very low and conditions for visual [gun] control poor. It is believed that no ships with radar control were in immediate area because no one fired until planes were visually detected.
>
> About 0745, three single engined dive bombers appeared out of direction of sun and all ships fired when visual contact was made. The three then separated and disappeared back in the clouds. One reappeared from the south, one from the northeast, and one from the north. These were all taken under fire by all ships. Two of these planes were splashed and the third, badly hit, crashed into a merchant ship.
>
> While this last action was taking place, another single engine dive bomber appeared from the west low over the land. This plane was also splashed when taken under fire by ships in the anchorage. We did not fire on this plane due to other ships being in line of fire.[15]

HARBOR CLEARANCE AT OKINAWA

> *A few days after our arrival in this theatre of operations we were visited by a "Val" of unfriendly character whose pilot chose to go in for a posthumous promotion and headed his plane for our ship. Many other ships fired on 'Val' but our Gunnery Officer is certain that more tracers from our gun crews hit him*

than any other ship's. On second thought 'Val' changed his mind. He probably thought we were too belligerent, but in any case he dove in the bay about 300 yards from our port quarter. A tense moment was followed by one of boundless enthusiasm. More cheering came from our men than from the most enthusiastic alumni at the football game.

—USS *Anchor* (ARS-13) War Diary, June 1945.

Photo 17-6

"Val" Navy Type 99 carrier bomber.
Japanese Operational Aircraft "Know Your Enemy!"
CinCPac - CinCPOA Bulletin 105-45

The salvage ship *Anchor* arrived at Naha Harbor, Okinawa, in the first week in June, tasked with raising and removing many of the eighty or ninety Japanese wrecks in the harbor, thereby freeing it for use by Allied shipping. A Stockton Colberg Boat Works product, she was the first Allied ship to proceed to the inner harbor, and believed to be the first since pre-war days to moor at a Japanese pier—tied up about 1,000 yards from the location Commodore Matthew C. Perry landed in 1853. Both as a symbol and in reality, to her crew the job given them marked a transition from destruction to construction.[16]

The voyage from Guam to Okinawa in convoy had been far from uneventful. Reports of a possible typhoon had started to come over the radio three days out from Guam. There were conflicting opinions about the location of the center of the storm, and its course and speed. But the general belief was that it would be safe to continue on the course over which the convoy had been routed. This was not so! At noon on the sixth day, the convoy received orders to reverse course. By 2200 the barometer had dropped to 29.44 inches of mercury, and was still falling at a rate of .4 hundredths of an inch per hour.[17]

Both the ships in convoy and of the escort force were directed to proceed independently, so as to lessen the risks of collision in the dark. *Anchor* rolled severely, as much as 55 degrees, so that a roll of 40 degrees was regarded, after a while, as comfortable. During the storm personnel of the deck force maintained a continuous watch in the holds to ensure that no gear broke loose; others watched the towing cable and its attached tow. The bridge team strove to keep the ship steady on course; radiomen were occupied receiving messages, using authenticators, creating headings, and sending out reports; and repair party personnel stood by to assist with pumps. By 0800 the following morning, the sea had become somewhat calmer.[18]

After some difficulties, the convoy reassembled and resumed a course for Okinawa, proceeding peacefully for the remainder of the voyage. Waiting on the pier to greet the salvage ship upon her arrival, was a large assembly of Marines, who had subsisted on Rations K and C for over two months. They came aboard with souvenirs of some value and interest, including Japanese rifles, swords, medals, chinaware, lacquerware and currency. Many of the men were anxious to trade items for Navy chow. Unfortunately, the food supply aboard was limited, and the ship was therefore, unable to accommodate the hungry Marines.[19]

Lt. Richard J. Cruise, and six men under him comprised the Salvage Crew aboard *Anchor*. The first wreck they tackled was a Japanese merchant ship, 300 feet in length and about 4,000 tons. It was raised in ten days, which was about half the time allowed for the job. Towed out to a "watery graveyard" well clear of the harbor entrance, several adroitly placed sticks of dynamite soon disposed of the wreck by creating flooding holes in it. Such work would continue until Naha Harbor was completely clear of ruined vessels, and shipping could enter it unhampered.[20]

RESCUE TUGS AT OKINAWA

At approximately 0905, sighted a formation of Japanese planes. Three of them left the formation and attacked this vessel simultaneously, while others headed for other ships on this station. Of the three which attacked this vessel, two were driven off, but the third effected a suicide bombing, crashing into the twin 40MM gun on fantail, and exploded. Fire broke out on fantail and spaces below.

While fighting fire and caring for wounded, this ship was attacked by two other Vals, who also attempted to suicide bomb. The first of these had been hit by a friendly [F6F] Hellcat [fighter aircraft], and was smoking while on suicide run. Was brought down by gunfire from forward guns, and crashed into water about 200 yards from ship. The other suicide bomber was hit by gunfire from this vessel, flew over ship, cutting off antenna, and crashed into water approximately 100 yards to starboard.

—Description of attacks on 16 April 1945 made by Japanese aircraft against the support landing craft USS *LCS(L)(3)-116* while she was assigned to a radar picket station off Okinawa.[21]

Colbert Boat Works-built *ATR-51* and *ATR-53* each earned battle stars for the Assault and Occupation of Okinawa. Information about their qualifying action(s) is scarce, but they were likely related to combat with Kamikaze planes, and/or being in the vicinity of, or sent to aid other ships damaged by suicide aircraft.

Photo 17-7

USS LCS(L)(3)-116 under way.
U.S. Navy photograph

For example, on the evening of 16 April (during her 1-28 April qualifying period for a star), *ATR-51* took the damaged landing craft USS *LCS(L)(3)-116* with dead on board under tow. The 158-foot vessel, armed with guns and rockets, had been hit by a Val earlier that day while assigned duty on a radar picket station.

On 30 June, the day of her qualifying action, *ATR-53* was in convoy, en route to, and a day out from, Okinawa. She arrived there at 1223 on 1 July. Her duties for the remainder of the month and beyond were the same as those of USS *Anchor*, diving on sunken ships and using explosive charges to clear them of obstructions to within 35 feet of the surface. *ATR-53*'s work was also at Naha Harbor, located on the southwest side of the island.[22]

18

Borneo Campaign

The objectives of the Allied Borneo Campaign of 1945 (1 May-15 August) were to deny Japan the continued fruits of its conquests in the Netherlands East Indies, and use of the approaches to those areas. To achieve these aims, an Australian-led force captured Tarakan Island to provide an airfield for support of an assault on Balikpapan, and seized Brunei Bay for an advanced fleet base that could protect resources in the area. The final phase was to occupy Balikpapan for its naval air and logistic facilities as well as its petroleum installations.[1]

Vice Adm. Daniel E. Barbey, USN (commander, 7th Amphibious Force) was designated commander, Balikpapan Attack Force. Rear Adm. Albert G. Noble, USN (commander, Amphibious Group Eight) was assigned as commander, Balikpapan Group. Rear Adm. Ralph S. Riggs, USN was assigned as commander, Cruiser Covering Group. A portion of the forces assigned in the Operations Order (OpOrder) for the assault of Balikpapan are identified in the following table.[2]

These include the Royal Australian Navy infantry landing ships *Manoora*, *Westralia*, and *Kanimbla*; thirty-nine USN YMSs and three steel-hulled AMs; the mine disposal support vessel *YP-421*; and assigned to the hydrographic unit, HMAS *Warrego* and *YMS-196*, as well as the net layer USS *Mango.*

Balikpapan Attack Group (TG 78.2): Rear Adm. Albert G. Noble, USN

Transport Unit (78.2.2): Capt. Allan P. Cousin, RANR(S)

Flagship HMAS *Manoora*, HMAS *Westralia*, HMAS *Kanimbla*

USS *Titania* (AKA-13)

USS *Carter Hall* (LSD-3)

Minesweeping Unit (78.2.9): Lt. Comdr. Thomas R. Fonick, USNR

Task Unit 78.2.91: Lt. Comdr. Thomas R. Fonick (CO, USS *Sentry*)

Flagship USS *Sentry* (AM-299), USS *Scout* (AM-296), USS *Scuffle* (AM-298)

Task Unit 78.2.92: Lt. Comdr. Blakeslee (CO, USS *Cofer*)

Flagship USS *Cofer* (APD-62) (high-speed transport carrying LCVPs)

YMSs originally assigned: USS YMSs **9*, *10*, *39*, *46*, *47*, *49*, *50*, *52*, *53*, *95*, *314*, *315*, *335*, *336*, *339*, *364*, *365*, *366*, *368*, *392* (*YMS-9* being held in reserve at Morotai)

YMSs reporting F-1 and F-Day: USS YMSs *65*, *84*, *224*, 269, *367*

USS *LSM-1* (medium landing ship serving as minesweep tender)

USCG *Gualala* (AOG-28) (gasoline tanker)
Service Unit (78.2.14): Lt. Comdr. Wallace
USS *YP-421*
Hydrographic Unit (78.2.17): Comdr. Colin Goyder Little, DSC RAN
Minesweeping sloop HMAS *Warrego* (U73)
USS *YMS-196*
USS *Mango* (AN-24)
Support and Covering Group (74.2): Rear Adm. Ralph S. Riggs, USN (Commonwealth units only)
Heavy cruiser HMAS *Shropshire* (D73)
Light cruiser HMAS *Hobart* (D63)
Cruiser HRMS *Tromp*
Destroyer HMAS *Arunta* (I30)[3]

The Dutch cruiser *Tromp* is included in the table, along with the RAN ships assigned to the Support and Covering Group, as a point of interest. One of the highest decorated Dutch warships of World War II, she was referred to as the "The Ghost Ship," because of the number of times the enemy erroneously claimed to have sunk her. Her crew preferred to call her "The Lucky Ship."

It's important to note that changes of ships were sometimes made as a result of them breaking down or being damaged or lost in combat. Though not listed in the OpOrder, Stockton minesweepers *YMS-95* and *97*, participated in and earned battle stars for operations at Balikpapan, for the periods identified in the table. *YMS-95* and other minesweepers also earned the Presidential Unit Citation.

Borneo Operations: Balikpapan Operation

USS YMS-95	15 Jun-9 Jul 45	Colberg Boat Works
USS *YMS-97*	27 Jun-15 Jul 45	Colberg Boat Works

MINESWEEPING AT BALIKPAPAN

Enemy shore battery fire was encountered by the YMS on numerous occasions from caves, pillboxes and gun emplacements. During F-13 [18 June], 12 [19 June], 10 [21 June], 9 [22 June], 8 [23 June], these batteries could be only temporarily silenced by the Cruiser Support Group since it was necessary for the cruisers to remain at ranges of 14[000] to 1600[0] yards. On FOX *minus seven [24 June] when sweeping had progressed sufficiently to permit destroyers to close to 6000 yards, only sporadic and inaccurate fire was then directed at the minesweepers.*

—Commander, Task Unit 78.2.9, Action Report, Minesweeping Unit Balikpapan, Borneo, NEI, 11 June to 1 July 1945, 11 July 1945.

Map 18-1

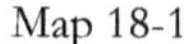

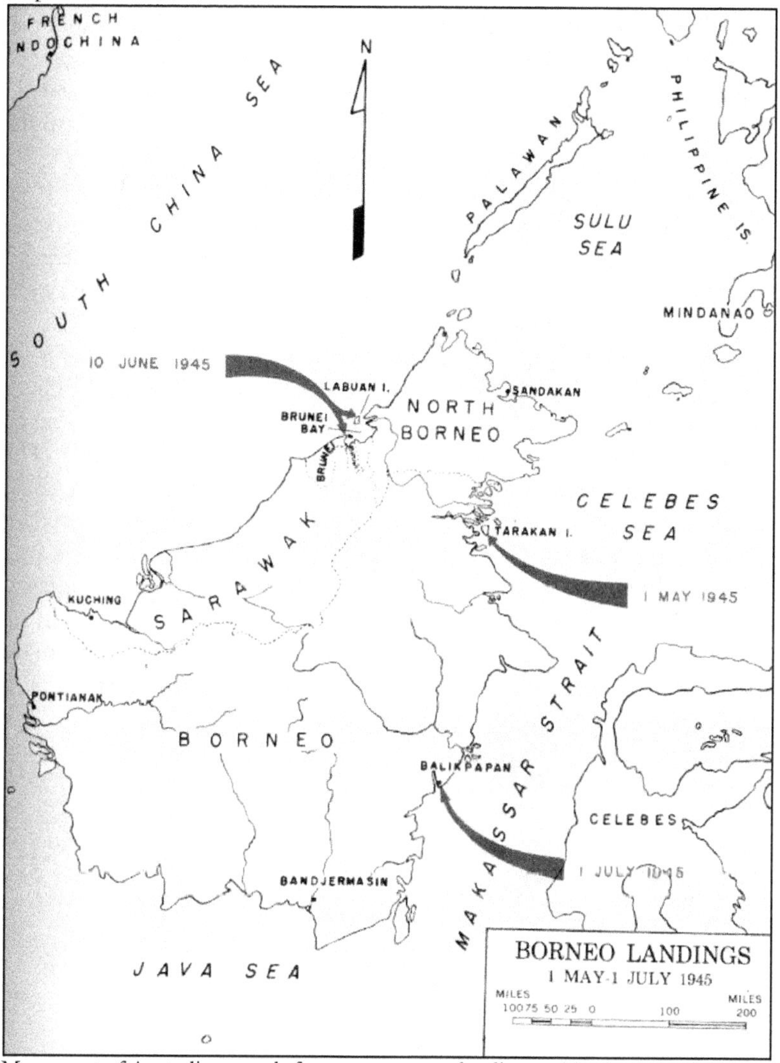

Movement of Australian assault forces en route to landings at Tarakan Island, at Labuan Island in Brunei Bay, and at Balikpapan; 1945 Borneo Campaign https://www.navy.gov.au/history/feature-histories/borneo-1945-amphibious-success-story

The mission of the minesweeping force at Balikpapan was to clear the channels, approaches, landing beaches and other designated areas of Allied and enemy mines prior to F-Day (1 July 1945) to allow safe passage for assault shipping. At 0642 on 15 June, YMSs of Task Unit 78.2.92 began due east of Balikpapan, a sweep of the approach track for moored and acoustic mines. Among this group of "sweepers" (newly arrived from Morotai Island in the Haimaheras) was Colberg Boat Works' *YMS-95*. Following astern of the YMSs in their swept waters, were the high-speed transport USS *Cofer* (carrying LCVP shallow-water minesweepers), gasoline tanker USCG *Gualala*, and the units of the Support and Covering Group.[4]

No mines were found that day, and none the next (16 June) when YMSs began moored, magnetic, and acoustic sweeps of the fire support areas, as minesweeping LCVPs (Higgins boats) concurrently commenced an exploratory sweep of reported Japanese minefields. Results were also negative on 17 June.[5]

At 1253 on 18 June, *YMS-50* exploded an Allied influence mine under her bow, and was seriously damaged. (Royal Australian Air Force PBY Catalinas had mined Balikpapan waters earlier in the war.) Seven minutes later, she came under fire from enemy shore batteries. The cruiser support group then commenced counter-battery fire. At 1312, LCVPs from the *Cofer* arrived to rescue personnel, and attempted to take the minesweeper in tow. They were taken under fire and abandoned the towing effort. All crewmembers were rescued, and the light cruiser USS *Denver* (CL-57) sank the hulk of *YMS-50* with gunfire.[6]

That evening, sweeping operations ceased at 1815 and night retirement, movement farther out to sea, commenced. At 1855, a "Pete" (Japanese Navy observation plane) closed the task unit. *Cofer* opened fire and as the aircraft sped away, the combat air patrol took up chase of the "snooper."[7]

Photo 18-1

Japanese Navy "Pete II" observation plane.
Japanese Operational Aircraft "Know Your Enemy!" CinCPOA Bulletin 105-45

The following morning, 19 June, the YMSs began a sweep of assault and close support areas. At 1100 they came under shore battery fire. No hits occurred and the support group silenced the batteries. The sweepers were taken under fire a second time at 1250, again there were no hits. At 1900, the YMSs recovered gear and commenced night retirement.[8]

On 20 June, the YMSs began a sweep of fire support areas, and mid-afternoon brought another mine strike. At 1322, *YMS-368* exploded an Allied influence mine near her stern, causing severe structural damage. *Cofer* personnel effected temporary repairs, and the minesweeper remained afloat.[9]

The following morning, 21 June, brought enemy shore battery fire during sweep operations in assault and close support areas, but no casualties. Mid-afternoon, *YMS-335* was struck by an enemy round on her 3-inch gun platform, resulting in four killed and five wounded. The commanding officer of *YMS-95*, Lt. (jg) George E. Newby Jr., USNR, described this, and ensuing action in a report on Balikpapan operations from 11 June to 2 July 1945:

> On southern pass along area, enemy shore batteries opened fire on unit. Unit retired and shore batteries taken under fire of Task Group 784.2. Regrouped unit and made a second attempt. *YMS 52* in the lead and *YMS 95* the second ship. These were the only vessels for this attempt. *YMS 52* was headed on an easterly course on north side of area and this vessel was just about to come to an easterly course, from the western boundary, when the shore batteries opened up on the *YMS 52*. This vessel was able to fire three rounds 3"/50 cal. ammunition before ship came about to course which would get us out of area. Lt.(jg) Reiser gunnery officer, and lookouts, reported that our second round made a direct hit on the shore battery. The range was 4400 yards. A plume of white smoke rose from the area which harbored the battery. Immediately Task Group 74.2 [Fire Support Units], took the batteries under their fire. The "O" type sweep gear of the M Mark 6(A) gear was cut, to enable this vessel to turn quicker and clear the area in quickest possible time. The gear was recovered for repairs.[10]

The next day, the 22nd, brought more of the same while working the same waters. At 1014 on the 22nd, *YMS-10* was hit in the bow by an exploding round which made a hole approximately three feet in diameter above the waterline, but caused no personnel casualties. Astern of *YMS-10* in a three-ship sweep formation, and ahead of *YMS-315*, *YMS-95* was able to get off four 3-inch rounds before a reverse in course, eliminated the effectiveness of the gun mount. These batteries were again taken under fire by Task Group 74.2 Fire Support Units. Aboard *YMS-95*, lit smoke pots were dropped over the stern and also placed on the deck aft, to form a smoke screen.[11]

Photo 18-2

One of two Japanese 120MM Naval Coast Defense Guns located at Waites Knoll, North of Manggar Airfield, which covered the beach and channel to Balikpapan. Australian War Memorial photograph 069501

Providence was smiling on the sweeps on 23 June. *YMS-368*, which had been damaged by a mine explosion three days earlier, left under tow by *YMS-335*, bound for Tawi, Tawi Island in the southwestern Sulu archipelago, Philippines, between the Celebes Sea and the Sulu Sea. *YMS-364* was struck by an enemy projectile which failed to explode after striking her bridge, passing through a few bulkheads, and landing in the captain's bunk. He chose not to keep it as a souvenir and had it thrown overboard.[12]

Twenty-four June was also a good day. As the YMSs carried out sweeping operations in the assault area, *Scout* (AM-296) with *YMS-6* conducted a sweep (against moored mines) of the approach track to Balikpapan to widen the existing swept channel. The sweeping unit came under sporadic fire that day, but no hits were scored.[13]

During operations in the assault lane and close support area the following day, 25 June, the leading YMS came under fire from shore batteries, but suffered no hits. *YMS-52* exploded an influence mine, as did *YMS-365* which damaged her sweep, and forced her to retire to effect repairs. *YMS-39* detonated an influence mine, and cut a moored mine, which was sunk by gunfire. That night, the task unit came under attack at 2030 by four enemy aircraft tentatively identified as "Peggy" Army heavy bombers. Three minutes later, *Cofer* was attacked by three planes, all of which dropped torpedoes which passed under her bow. Their remaining service to Japan was short-lived. One was shot down by *Cofer*, another by *Cofer* in conjunction with other vessels of Task Unit 78.2.9, and the final one by *Sentry*.[14]

YMS-95's commanding officer described these events:

> At 20:30 25th June, condition red was flashed, and this vessel immediately went to General Quarters. At 20:35, an enemy bomber (Japanese) was caught in anti-aircraft fire from approximately six different ships of our unit. It burst into flames, and crashed into the water about two hundred yards ahead of this vessel. No survivors were seen. At 20:40, another Japanese bomber came in low over the water, from the beach. It passed approximately three hundred feet over the stern. This vessel opened up on it with twenty millimeter and fifty cal. fire. Hits were made on the tail and fuselage. Other ships picked it up in their fire as it continued on its course. The last glimpse of it, was when it was trying to gain altitude. At 20:48 an explosion appeared at approximately where the Japanese air strip had been on the beach, to the east of Balikpapan point. Many small fires were visible about the entire area. For the remainder of the night, this vessel stayed at general quarters.[15]

Photo 18-3

Japanese "Peggy" Army heavy bomber.
Japanese Operational Aircraft "Know Your Enemy!" CinCPOA Bulletin 105-45

YMS-365 AND *YMS-39* SUNK WITH CASUALTIES

Two of the four YMSs lost at Balikpapan occurred on 26 June, when the sweepers encountered both moored and influence mines in the assault area. At 1008 an influence mine was exploded by a magnetic sweep with no damage to the towing ship. Also, that day, three cut moored mines were destroyed by gunfire from the destroyer USS *Stevens* (DD-479).[16]

At 1343, *YMS-365* detonated an influence mine with her magnetic sweep. However, less than an hour later, one of these types of mines exploded directly beneath her, breaking her keel amidships. *YMS-364* removed surviving crew, and *YMS-196* then opened fire on the hulk and sank the bow. The stern stubbornly remained afloat until 1645. A short time earlier, what was believed to be an Allied magnetic mine exploded under *YMS-39*. As her superstructure disintegrated, the vessel capsized and sank in less than one minute. Two additional mine explosions then occurred in short succession.[17]

After *YMS-196* and LCVPs from *Cofer* (APD-62) and *Schmitt* (APD-76) rescued survivors from *YMS-365* and *YMS-39*, the unit ceased sweeping and proceeded to anchorage at 1855. It had been a disastrous day for them. The only bright spot was that fire support station waters were sufficiently cleared for destroyers to move in and render close fire support to minesweepers engaged in sweeping the assault area. Further enemy gunfire directed against the unit was sporadic and inaccurate.[18]

Photo 18-4

Minesweeper USS *YMS-39* under way, location and date unknown.
Courtesy of City Island Nautical Museum

ADDITIONAL DAMAGE WROUGHT BY MINES

On 28 June, *YMS-47* exploded an influence mine under her, which blew a large hole in her stern and flooded her after compartments. *YMS-366* and *YMS-49*, assisted by LCVPs from *Cofer* (APD-62), *Kline* (APD-120), and *Schmitt* (APD-76), rescued all personnel and succeeded in towing *YMS-47* clear of the minefield. She was taken alongside *Scout* for the purpose of pumping and emergency repairs. Although her after deck was awash, once the engine room compartments were emptied of water and, patched up, she stayed afloat.[19]

Photo 18-5

A vehicle personnel landing craft (LCVP) off Balikpapan, Borneo, 1 July 1945. Australian War Memorial photograph 110411

On 30 June, *YMS-314* exploded an Allied influence mine about 100 yards astern, resulting in only minor damage.[20]

YMS Ship and Personnel Casualty Summary

		Officers			**Enlisted**		
Date	**Ship**	**KIA**	**WIA**	**MIA**	**KIA**	**WIA**	**MIA**
18 June	*YMS-50* (sunk)		2			9	
20 June	*YMS-368* (damaged)					3	
21 June	*YMS-335* (damaged)		1		4	4	
22 June	*YMS-10* (damaged)						
23 June	*YMS-364* (damaged)						
26 June	*YMS-365* (sunk)		2			4	
26 June	*YMS-39* (sunk)				3	6	1
28 June	*YMS-47* (damaged)					10	
28 June	*YMS-49* (damaged)						
30 June	*YMS-314* (damaged)						
9 July	*YMS-84* (sunk)						
	Total		5		7	36	1

Notes: (A) One enlisted man aboard *YMS-366* went missing at sea, not as a result of enemy action.

(B) *YMS-365* also damaged earlier on 23 June.[21]

By 1 July, the date of the landings at Balikpapan, the mine force had swept a total of fifteen Allied influence and thirteen enemy moored contact mines. The Japanese chemical horn moored contact mines were reported as Type 6 and Type 93. Mines cut were destroyed by machine gun fire; only one exploded when struck by rifle fire. Allied influence mines proved most dangerous as explained in a brief report by commander, Task Unit 78.2.9:

> This operation was extremely hazardous from the minesweeping point of view. Evidence is afforded by the high rate of casualties to YMS caused by mine explosions. Sweeping allied influence mines is admittedly slow, difficult and dangerous in water less than 5 fathoms. Most of the allied mines swept at Balikpapan were in 2½ to 4 fathoms. The danger to the sweeping vessel in this particular area was greatly increased by enemy shore fire and contact mines.[22]

YMS-84 was sunk by a moored contact mine on 9 July; the last surface ship of the U.S. Navy to be lost in that far corner of the Pacific where the realm of the Golden Dragon sweeps down under the Southern Cross.[23]

Photo 18-6

Japanese soldiers inspect a mine in shallow water at Balikpapan, Borneo, 17 June 1942. Australian War Memorial photograph

SACRIFICES LEAD TO SUCCESSFUL LANDINGS

The bravery of the crewmen of the "sweeps" at Balikpapan was particularly notable because it came on the heels of heavy losses and damage to minesweepers by mines and Japanese artillery fire at Tarakan Island and Brunei Bay (not part of this book). Due to the perseverance of the American minesweepers and Underwater Demolition Team 11, the amphibious landing was made at Balikpapan on 1 July as scheduled.[24]

After an intense two-hour bombardment, assault waves moved ashore, landing without a single casualty in spite of enemy artillery, mortar and small arms fire. The troops met with increased resistance as they progressed inland. Fire support was provided by cruisers day and night, and continued through 7 July. A subsequent landing was made at Cape Penajam on 2 July without casualties. However, minesweepers paid a heavy price off Balikpapan.[25]

During the Borneo Campaign, the U.S. Pacific Mine Force lost *Salute* (AM-294), *YMS-39*, *50*, *84*, *365*, and *481*, and suffered damage to twelve others—*YMS-10*, *47*, *49*, *51*, *314*, *329*, *334*, *335*, *339*, *363*, *364*, and *368*.[26]

USN UNIT CITATIONS AND RAN BATTLE HONOURS

The below listed U.S. Navy minesweepers were awarded the Presidential Unit Citation, for heroic actions at Balikpapan.

Presidential Unit Citation (Balikpapan, Borneo, 15 June-1 July 1945)		
USS *Scout* (AM-296)	USS *YMS-49*	USS *YMS-335*
USS *Scuffle* (AM 298)	USS *YMS-50*	USS *YMS-336*
USS *Sentry* (AM-299)	USS *YMS-52*	USS *YMS-339*
USS *YMS-9*	USS *YMS-53*	USS *YMS-364*
USS *YMS-10*	USS *YMS-95*	USS *YMS-365*
USS *YMS-39*	USS *YMS-196*	USS *YMS-366*
USS *YMS-46*	USS *YMS-314*	USS *YMS-368*
USS *YMS-47*	USS *YMS-315*	USS *YMS-392*

Note: These minesweepers, and others involved in the Borneo Campaign, also earned battle stars, including YMS-97.

The following Royal Australian Navy ships and Beach Commando earned Battle Honour BORNEO 1945:

- Heavy cruiser HMAS *Shropshire* (D73)
- Light cruiser HMAS *Hobart* (D63)
- Destroyers HMAS *Arunta* (I30) and *Warramunga* (I44)
- Infantry landing ships HMAS *Kanimbla*, *Manoora*, and *Westralia*

- Frigates HMAS *Barcoo* (K375), *Burdekin* (K376), HMAS *Gascoyne* (K354), *Hawkesbury* (K363), and *Lachlan* (K364)
- Minesweeping sloop HMAS *Warrego* (U73)
- Minesweeping corvettes HMAS *Latrobe* (J234) and *Stawell* (J348)
- RAN Beach Commando

Photo 18-7

Painting by Dennis Adams of the "Quartermaster of the *Warrego*" at the wheel, while in the chart room behind him, the officer of the watch checks the chart. Australian War Memorial photograph ART 22363

Postscript

On 15 August [1945] President Truman was able to announce the unconditional surrender of the Japanese Imperial Government. Orders were sent out for the occupation of Japan and Korea.... Overnight [there was] a change in the attitude of the men in the combat zone from one of war to one of peace. Suddenly everyone wanted to go home. No one was interested in becoming a member of the occupational forces.

—Excerpts from a description by Vice Adm. Daniel E. Barbey, USN, (Ret.) in *MacArthur's Navy* about the perspective of Sailors and Marines in the Pacific following the end of World War II, which could also describe the attitude of servicemen in Europe after the cessation of fighting there.[1]

Photo Postscript-1

U.S. Navy sailors celebrating VJ (Victory over Japan) Day at sea, 15 August 1945. Naval History and Heritage Command photograph #UA 570.61

Photo Postscript-2

Empty beer bottles strewn along Mindil Beach, Darwin, 16 August 1945, the day after Army soldiers celebrated Victory over Japan (VJ) Day.
Australian War Memorial photograph P04287.006

On 15 August 1945, Adm William F. Halsey, commander Third Fleet, received orders from Fleet Admiral Chester W. Nimitz, commander in chief, Pacific Fleet, to cease all operations against Japan. The cease fire order was received too late to stop the first of the day's air strikes against Japan, but the second strike which had been launched, was recalled in time. Two weeks later, Nimitz arrived in Tokyo Bay aboard a PB2Y Coronado seaplane, in preparation for Japan's formal surrender aboard the battleship *Missouri* (BB-63).

SURRENDER CEREMONY IN TOKYO BAY

It is my earnest hope, and indeed the hope of all mankind, that from this solemn occasion a better world shall emerge out of the blood and carnage of the past—a world dedicated to the dignity of man and the fulfillment of his most cherished wish for freedom, tolerance and justice.

—Closing remark made by Gen. Douglas MacArthur during a speech at the Surrender Ceremony of Japan aboard the battleship USS *Missouri* (BB-63) on 2 September 1945.[2]

Photo Postscript-3

Fleet Admiral Chester W. Nimitz, commander in chief, Pacific, and Pacific Ocean Areas, signs the Instrument of Surrender as United States representative, 2 September 1945. Naval History and Heritage Command photograph #NH 58082

Present in Tokyo Bay, on 2 September, were 255 Allied naval ships and merchant vessels. None of these were Stockton vessels, but four of her minesweepers were soon engaged in dangerous, post-war work in Japanese waters, conducting mine clearance operations.[3]

THIRD FLEET OPERATIONS AGAINST JAPAN

Extensive Third Fleet Operations against the Japanese empire in July and early August 1945, had finished Japan as a war-making nation, in spite of its four million men still under arms. With Allied offensive operations against Japan ongoing, mine clearance operations were conducted in the East China Sea. This sea, which laps against the eastern coast of China, extends eastward to the Ryukyu Islands; north to Kyushu, the southernmost of Japan's main islands; and northwest to Cheju Island off South Korea.[4]

Japanese minefields stretched from north of Formosa to Kyushu, barring entrance to the East China Sea and to its north, the Yellow Sea. Among U.S. Navy mine forces involved in clearing Area JUNEAU—9,000 square miles enclosing a suspected enemy mine barrier—was Sweep Unit Five (Task Unit 39.11.5):

- Steel-hulled minesweepers *Pochard* (AM-375), *Requisite* (AM-109), *Token* (AM-126), *Tumult* (AM-127), *Velocity* (AM-128), *Zeal* (AM-131)
- Wooden-hulled minesweepers *YMS-362*, *YMS-388*, *YMS-434*
- Motor gunboats *PGM-9*, *PGM-20*[5]

Colbert Boat Works-built *YMS-388* participated in, and earned a battle star for this operation.

3rd Fleet Operations Against Japan		
USS *YMS-388*	5-31 Jul 45	Colberg Boat Works

MINESWEEPING OPERATIONS PACIFIC

> *The mine clearance plan for the Western Pacific, set up by the [commander,] Minecraft [Pacific Fleet] staff, covered several phases. First and most pressing task was to expedite evacuations of Allied prisoners of war in the Japanese empire, and entry of occupational troops. That meant immediate clearance of channels through enemy minefields and U.S. laid influence minefields into key Japanese ports. Next was the clearance of additional channels and harbor facilities to allow safe entry of the supplies necessary to support occupational troops. Then came the sweeping of sea lanes around south Kyushu and between the various occupied Japanese ports so U.S. ships could move freely about their duties. Such operations, termed "occupational sweeping," covered clearances in 14 different areas spread over 17,000 square miles.*
>
> —Arnold S. Lott in *Most Dangerous Sea.*[6]

Photo Postscript-4

USS *YMS-97* in San Francisco Bay, 1945-1946.
Naval History and Heritage Command photograph NH 81376

Four minesweepers built by Colberg Boat Works—*YMS-97*, *YMS-386*, *YMS-387*, *YMS-388*—operated in Asian waters until 2 March 1946 as part of the overall mine clearance effort. Pursuing their dangerous craft as during the war, they were at least serene in the thought that they would not be attacked by hostile aircraft, vessels or submarines. These were the last stalwart Stockton ships to earn battle stars, these coming in the aftermath of World War II, while working waters during the periods and in the areas identified in the following tables. The "(No BS)" annotation following the names of some ships, means that they did not receive battle stars for those particular efforts.

Special Operations: Minesweeping Operations Pacific: Nagoya (Honshu Area)

USS *YMS-386*	28 Sep-25 Oct 45	Colberg Boat Works
USS *YMS-388*	7-25 Oct 45	Colberg Boat Works

Special Operations: Minesweeping Operations Pacific

USS *YMS-97*	14 Oct 45-2 Mar 46	Colberg Boat Works

Special Operations: Minesweeping Operations Pacific: Fukuoka (Kyushu-Korea Area)

USS *YMS-387*	9-28 Nov 45	Colberg Boat Works
USS *YMS-388* (No BS)	6 Nov-26 Dec 45	Colberg Boat Works

Special Operations: Minesweeping Operations Pacific: Fusan (Kyushu-Korea Area)

USS *YMS-388* (No BS)	13-28 Jan 46	Colberg Boat Works

Special Operations: Minesweeping Operations Pacific: Kobe (Honshu Area)

USS *YMS-386* (No BS)	20 Dec 45-2 Mar 46	Colberg Boat Works
USS *YMS-387* (No BS)	10 Dec 45-2 Mar 46	Colberg Boat Works

Special Operations: Minesweeping Operations Pacific: Yangtze River Approaches

USS *YMS-386* (No BS)	22 Feb-2 Mar 46	Colberg Boat Works
USS *YMS-387* (No BS)	22 Feb-2 Mar 46	Colberg Boat Works
USS *YMS-388* (No BS)	22 Feb-2 Mar 46	Colberg Boat Works

Photo Postscript-5

USS *YMS-387* at Navy Section Base Santa Barbara, California, 20 November 1943. MCAS Santa Barbara photograph BARB-550-10-43

Appendix A: Stockton-built Military Vessels

COLBERG BOAT WORKS

U.S. Navy 183-Foot Rescue and Salvage Ships (3)

Vessel	Comm.	Vessel	Comm.
Anchor (ARS 13)★	23 Oct 43	*Extractor* (ARS 15)	3 Mar 44
Protector (ARS 14)	28 Dec 43		

★ World War II Campaign Ribbon Battle Star

U.S. Navy 172-Foot Ocean Minesweepers (3)
(AM designation later changed to MSO)

Vessel	Comm.	Vessel	Comm.
Dynamic (AM 432)	15 Dec 53	*Embattle* (AM 434)	16 Nov 54
Engage (AM 433)	29 Jun 54		

Campaign and Unit Awards

Dynamic (MSO 432)	*Engage* (MSO 433)	*Embattle* (MSO 434)
AE 21 Feb-7 Mar 62 I	AE 28 Aug-6 Sep 60 G	AE 30 Mar-18 Apr 65
26 Apr-14 May 62 I	25 May-2 Jun 62 G	4 May-1 Jun 65
23 May-14 Jun 62 I	4 Jul-1 Aug 62 I	24 Jun-3 Jul 65
CR 29 Aug-15 Oct 70	11-13 Aug 62 G	MU 11 Nov 66-6 May 67
MU 1 Oct 65-30 Jun 66	6-29 Sep 64 I	5 May-7 Sep 68
1 Sep 67-31 May 68	HS 5-10 Jun 80 15	RG 6 Jan-6 Feb 70
RG 12 Jan-10 Feb 68	RG 12 Apr-7 May 66	27 Feb-18 Mar 70
9 Mar-17 Apr 68	VS 2 Apr-7 May 66	VS 4-30 Jul 65
23 Apr-9 May 68	11 Jan-3 Oct 71	6 May-6 Jun 68
19 Apr-19 May 69	1 Nov-4 Dec 72	10 Jul-25 Aug 68
16-31 Aug 70	30 Jan 73	4 Jan-10 Feb 70
20 Oct-4 Nov 70		26 Feb-21 Mar 70
VS 15 Nov-65-25 Jan 66		
4 Mar-6 Jul 66		
7 May-15 Jun 66		

Legend:

AE: Armed Forces Expeditionary Medal
G: Quemcy-Matsu (23 Jul 58-1 Jun 63)
I: Vietnam (1 Jul 58-3 Jul 65)
CR: Combat Action Ribbon
HS: Humanitarian Service Medal
15: Cuban Resettlement (27 Apr 80 - Still Open)
MU: Meritorious Unit Commendation
RG: Republic of Vietnam Meritorious Unit Citation – Gallantry
VS: Vietnam Service Medal

U.S. Navy 165-Foot Rescue Tugs (4)

Vessel	Comm.	Vessel	Comm.
ATR 50	5 May 44	*ATR 52*★	31 Jul 44
ATR 51★★	22 Jun 44	*ATR 53*★	11 Sep 44

U.S. Navy 136-Foot Yard Minesweepers (9)

Vessel	Delivery	Vessel	Delivery
YMS-95★★ PUC	12 Mar 42	*YMS-385*	30 Apr 43
YMS-97★★★	11 Apr 42	*YMS-386*★	23 Jul 43
YMS-99★	27 Apr 42	*YMS-387*★	17 Aug 43
YMS-383★	24 Feb 43	*YMS-388*★★★★	24 Sep 43
YMS-384	9 Apr 43		

PUC: Presidential Unit Citation

U.S. Navy 136-Foot Patrol Craft Sweepers (3)

Vessel	Comm.	Vessel	Comm.
PCS 1402★★★	13 Jan 44	*PCS 1404*★★★	30 Mar 44
PCS 1403★★★★	17 Feb 44		

U.S. Navy 120-Foot Yard Patrol Craft (2)

Vessel	Delivery	Vessel	Delivery
YP 645	1945	*YP 646*	1945

U.S. Army 45-foot Barges (2)

Vessel	Built	Vessel	Built
Barge *37*	1943	Barge *38*	1943

U.S. Army 40-foot Launch (1)

Vessel	Built
McCloud	1942

HICKINBOTHAM BROTHERS

U.S. Army 176-Foot Coastal Freighters (8)

Vessel	Built	Disposition
FS 404	10/44-4/45	USCG-manned, to the Philippines as *Venus*
FS 405	10/44-4/45	USCG-manned, to the Philippines as *Vizcaya*
FS 406	10/44-4/45	USCG-manned, lost in typhoon off Okinawa 1945
FS 407	10/44-4/45	USCG-manned, to USN 1950 as *AKL 31*
FS 408	10/44-4/45	USCG-manned
FS 409	10/44-4/45	USCG-manned
FS 410	10/44-4/45	USCG-manned, lost in typhoon off Okinawa 1945
FS 411	10/44-4/45	USCG-manned, to USN 1950 as *AG 177*, reefed 2015

U.S. Army 110-Foot Derrick Barges (7)

Vessel	Built	Disposition	Vessel	Built
BD 482	3/43-4/43	Sold as *Beaver*	*BD 800*	6/43-9/43
BD 483	3/43-4/43	Sold as *Shelby*	*BD 801*	6/43-9/43
BD ?	6/43-9/43	Sold as *Hercules*	*BD 802*	6/43-9/43
BD 799	6/43-9/43			

U.S. Army 74-Foot Small Tugs (5)

Vessel	Built	Disposition
ST 146	6/43-10/43	
ST 147	6/43-10/43	
ST 148	6/43-10/43	
ST 149	6/43-10/43	To Portugal 1953 as *Monte Grande*
Hickinbotham I	1944	Built for in-house use

U.S. Army 52-Foot SP Balloon Barges (2)

Vessel	Built	Vessel	Built
BSP 1621	Sep 43	BSP 1622	Sep 43

U.S. Army 52-Foot Balloon Barges (10)

Vessel	Built	Vessel	Built
BB 1623	9/43-10/43	*BB 1628*	9/43-10/43
BB 1624	9/43-10/43	*BB 1629*	9/43-10/43
BB 1625	9/43-10/43	*BB 1630*	9/43-10/43
BB 1626	9/43-10/43	*BB 1631*	9/43-10/43
BB 1627	9/43-10/43	*BB 1632*	9/43-10/43

U.S. Army 75-Foot Balloon Barges (10)

Vessel	Built	Vessel	Built
BB 1633	9/43-10/43	*BB 1638*	9/43-10/43
BB 1634	9/43-10/43	*BB 1639*	9/43-10/43
BB 1635	9/43-10/43	*BB 1640*	9/43-10/43
BB 1636	9/43-10/43	*BB 1641*	9/43-10/43
BB 1637	9/43-10/43	*BB 1642*	9/43-10/43

U.S. 62-Foot Army Tenders (2)

Vessel	Built	Vessel	Built
Tender	1944	Tender	1944

U.S. Army 50-Foot Tank Landing Barges (22)

Vessel	Built	Vessel	Built
BTL 455	12/42-1/43	*BTL 466*	12/42-1/43
BTL 456	12/42-1/43	*BTL 467*	12/42-1/43
BTL 457	12/42-1/43	*BTL 468*	12/42-1/43
BTL 458	12/42-1/43	*BTL 469*	12/42-1/43
BTL 459	12/42-1/43	*BTL 470*	12/42-1/43
BTL 460	12/42-1/43	*BTL 471*	12/42-1/43
BTL 461	12/42-1/43	*BTL 472*	12/42-1/43
BTL 462	12/42-1/43	*BTL 473*	12/42-1/43
BTL 463	12/42-1/43	*BTL 474*	12/42-1/43
BTL 464	12/42-1/43	*BTL 546*	12/42-1/43
BTL 465	12/42-1/43	*BTL 547*	12/42-1/43

GUNTERT & ZIMMERMAN

U.S. Army 140-Foot Derrick Barges (2)

Vessel	Built	Disposition
BD 6646	1954	To USN as YD 218, sold 1967
BD 6652	1954	To USN as YD 222, reclassified as IX 536 2002

KYLE & COMPANY

U.S. Army 162-Foot Coastal Tankers (11)

Vessel	Built	Disposition
Y 28	1/44-6/44	
Y 29	1/44-6/44	Sold 1946 as *J. J. Kelly*
Y 30	1/44-6/44	Sold 1946 as *Argo*
Y 31	1/44-6/44	To Norway 1946 as *Liten*
Y 32	1/44-6/44	To Honduras 1946 as *Ana*
Y 33	1/44-6/44	To Norway 1946 as *Angelus*
Y 34	1/44-6/44	To Honduras 1946 as *Tania*
Y 35	1/44-6/44	To the Philippines 1946 as *Y 35*
Y 44	7/44-9/44	
Y 45	7/44-9/44	To the Philippines 1946 as *Y 45*
Y 46	7/44-9/44	To USAF 1946 as *C-50-1171*

U.S. Army 120-Foot Gasoline Barges (8)

Vessel	Built	Disposition	Vessel	Built	Disposition
BG 1182	4/43		*BG 1186*	4/43	
BG 1183	4/43	Sold as *P. S. No. 76*	*BG 1187*	4/43	
BG 1184	4/43	Sold as *ST 20*	*BG 1188*	4/43	
BG 1185	4/43	Sold as *Manson 45*	*BG 1189*	4/43	

U.S. Army 110-Foot Deck Barges (14)

Vessel	Built	Vessel	Built
BC 522	9/42-12/42	*BC 529*	9/42-12/42
BC 523	9/42-12/42	*BC 530*	9/42-12/42
BC 524	9/42-12/42	*BC 531*	9/42-12/42
BC 525	9/42-12/42	*BC 532*	9/42-12/42
BC 526	9/42-12/42	*BC 533*	9/42-12/42
BC 527	9/42-12/42	*BC 534*	9/42-12/42
BC 528	9/42-12/42	*BC 535*	9/42-12/42

U.S. Navy 110-Foot Open Lighter (1)

Vessel	Built
YC 1419	1952

U.S. Navy 110-Foot Closed Lighter (3)

Vessel	Built	Disposition
YFN 1163	1952	
YFN 1191	1952	Scrapped 1996
YFNX 30	1952	Sold to Panama 2007 as *Discovery*

U.S. Navy 163-Foot Tank Barge (1)

Vessel	Built	Disposition
YO 231	1943	Disposed of 1945

MOORE EQUIPMENT COMPANY

U.S. Navy 104-Foot Seaplane Derricks (12)

Vessel No.	Built	Disposition
YSD 35	1944	Lost 1946
YSD 36	1944	Lost off Okinawa 1946
YSD 37	1944	Lost off Eniwetok 1946
YSD 42	1944	Lost off Guam 1980
YSD 43	1944	Lost off Eniwetok 1946
YSD 44	1944	Scrapped 1972
YSD 45	1944	Destroyed 1947
YSD 46	1944	To NDRF 1974
YSD 47	1944	Sold 1960
YSD 48	1944	Lost to typhoon Louise 1945
YSD 49	1944	Scrapped 1971
YSD 50	1945	Sold 1971, later MPE 50

U.S. Navy 110-Foot Open Lighters (4)

Vessel No.	Delivery Date	Vessel No.	Delivery Date
YC 1089	1945	*YC 1091*	1945
YC 1090	1945	*YC 1092*	1945

U.S. Navy 110-Foot Covered Lighters (6)

Vessel No.	Delivery Date	Vessel No.	Delivery Date
YFN 814	1943	*YFN 817*	1943
YFN 815	1943	*YFN 818*	1943
YFN 816	1943	*YFN 819*	1943

U.S. Navy Landing Craft (2)

Vessel	Deliver Date	Quantity Built
56-Foot LCM(6)	1943-44	Uncertain
50-Foot LCM(3)	1943-44	Uncertain

POLLOCK-STOCKTON SHIPBUILDING

U.S. Navy 194-Foot Net Layers (10)

Vessel	Comm.	Vessel	Comm.
Lancewood (AN 48)★	18 Oct 43	*Spicewood* (AN 53)★	7 Apr 44
Papaya (AN 49)★★	1 Dec 43	*Manchineel* (AN-54)	26 Apr 44
Cinnamon (AN-50)★	10 Jan 44	*Torchwood* (AN 55)	12 May 44
Silverbell (AN 51)★	16 Feb 44	*Winterberry* (AN 56)★	30 May 44
Snowbell (AN 52)★	16 Mar 44	*Viburnum* (AN 57)	27 Jun 44

★ World War II Campaign Ribbon Battle Star

U.S. Navy 260-Foot Lighters (29)

Vessel	Built	Vessel	Built
YFN 619	1943	YFN 1003	1945
YFN 620	1943	YFN 1004	1945
YFN 621	1943	YFN 1005	1945
YFN 622	1943	YFN 1006	1945
YFN 623	1943	YFN 1007	1945
YFN 738	1943	YFN 1008	1945
YFN 739	1943	YFN 1009	1945
YFN 740	1943	YFN 1010	1945
YFN 741	1943	YFN 1011	1945
YFN 742	1943	YFN 1012	1945
YFN 998	1945	YFN 1013	1945
YFN 999	1945	YFN 1014	1945
YFN 1000	1945	YFN 1015	1945
YFN 1001	1945	YFN 1016	1945
YFN 1002	1945		

U.S. Navy 261-Foot Barracks Barges (4)

Vessel	Built	Vessel	Built
APL 23	1945	*APL 25*	1945
APL 24	1945	*APL 26*	1945

U.S. Navy 528-Foot Drydock (1)

Vessel	Built
Steadfast (YFD 71)	

U.S. Navy 93-Foot Drydock Sections (4)

Vessel	Built	Vessel	Built
ABSD 3A	May 44	*Los Alamos* (ABSD 7E)	Mar 45
ABSD 3B	May 44	*Los Alamos* (ABSD 7F)	Mar 45

U.S. Navy 150-Foot Repair Barges (2)

Vessel	Built	Vessel	Built
YR 67	15 Sep 44	*YR 68*	10 Nov 44

STEPHENS BROTHERS

South Vietnamese 152-Foot Coastal Minesweepers (3)

Vessel	Built	Vessel	Built
Ham Tu (MSC 281)	26 Jun 59	*Bach Dang II* (MSC 283)	18 Sep 59
Chung Duong II (MSC 282)	21 Aug 59		

French 144-Foot Coastal Minesweepers (13)

Vessel	Built	Vessel	Built
Begonia (MSC 83)	22 Jul 53	*Pivoine* (MSC 125)	28 Aug 54
Coquelicot (MSC 84)	5 Oct 53	*Reseda* (MSC 126)	6 Nov 54
Giroflee (MSC 85)	30 Nov 53	*Tsushima* (MSC 255)	15 Jul 55
Laurier (MSC 86)	24 Jan 54	*Toshima* (MSC 258)	12 Aug 55
Magnolia (MSC 87)	26 Mar 54	*Gronsund* (MSC 256)	12 Mar 56
Myositis (MSC 123)	1 May 54	*Gulborgsund* (MSC 257)	11 Nov 56
Pavot (MSC 124)	12 Jun 54		

U.S. Navy 136-Foot Yard Minesweepers (3)

Vessel	Built	Vessel	Built
YMS 94★	11 Mar 42	*YMS 98*	23 Apr 42
YMS 96★★	11 Apr 42		

★ World War II Campaign Ribbon Battle Star

U.S. Navy 81-Foot Yard Patrol Craft (13)

Vessel	Built	Vessel	Built
YP 654	14 Mar 58	*YP 661*	1958
YP 655	20 Apr 58	*YP 662*	1958
YP 656	15 May 58	*YP 663*	15 Nov 58
YP 657	26 Jun 58	*YP 666*	1966
YP 658	28 Jun 58	*YP 667*	1967
YP 659	1958	*YP 668*	1967
YP 660	1958		

U.S. Navy 52-Foot Yard Patrol Craft (1)

Vessel	Built
YP 111 (ex-*Adventuress*)	25 Oct 40

U.S. Navy 52-Foot Cabin Cruisers (2)

Vessel	Built	Vessel	Built
No name	22 Aug 62	No name	22 Aug 62

U.S. Navy 41-Foot Ambulance Boat (1)

Vessel	Built
YH 4 (ex-*Sea Breeze*)★	1939

U.S. Coast Guard 38-Foot Picket Boats (20)

Vessel	Delivery	Vessel	Delivery
CG 4353/38367	12 Mar 42	*CG 4363/38377*	15 Apr 42
CG 4354/38368	18 Mar 42	*CG 4364/38378*	15 Apr 42
CG 4355/38369	25 Mar 42	*CG 4365/38379*	15 Apr 42
CG 4356/38370	24 Mar 42	*CG 4366/38380*	15 Apr 42
CG 4357/38371	24 Mar 42	*CG 4367/38381*	15 Apr 42
CG 4358/38372	16 Apr 42	*CG 4368/38382*	15 Apr 42
CG 4359/38373	15 Apr 42	*CG 4369/38383*	15 Apr 42
CG 4360/38374	15 Apr 42	*CG 4370/38384*	15 Apr 42
CG 4361/38375	15 Apr 42	*CG 4371/38385*	15 Apr 42
CG 4362/38376	15 Apr 42	*CG 4372/38386*	15 Apr 42

U.S. Army 104-Foot Rescue Boats (21)

Vessel	Delivery	Vessel	Delivery
P 110	3 Jun 42	*P 146*	1943
P 111	16 Jun 42	*P 209*	27 Oct 42
P 112	3 Jul 42	*P 210*	28 Oct 42
P 113	21 Jun 42	*P 211*	23 Oct 42
P 114	18 Jul 42	*P 280*	12 Feb 43
P 115	21 Jul 42	*P 281*	16 Feb 43
P 141	24 Jul 42	*P 282*	12 Mar 43
P 142	28 Jul 42	*P 283*	19 Mar 43
P 143	7 Nov 42	*P 284*	20 Mar 43
P 144	9 Nov 42	*P 285*	31 Mar 43
P 145	12 Dec 42		

U.S. Army 63-Foot Rescue Boats (33)

Vessel	Delivery	Vessel	Delivery
Q 679	6 Jun 44	*Q 696*	29 Sep 44
Q 680	16 Jun 44	*Q 697*	30 Sep 44
Q 681	24 Jun 44	*Q 698*	12 Oct 44
Q 682	30 Jun 44	*Q 699*	14 Oct 44
Q 683	7 Jul 44	*Q 700*	20 Oct 44
Q 684	12 Jul 44	*Q 701*	21 Oct 44
Q 685	22 Jul 44	*Q 702*	2 Nov 44
Q 686	28 Jul 44	*Q 703*	7 Nov 44

Q 687	1 Aug 44	*Q 704*	15 Nov 44
Q 688	11 Aug 44	*Q 705*	18 Nov 44
Q 689	22 Aug 44	*Q 706*	25 Nov 44
Q 690	6 Sep 44	*Q 707*	4 Dec 44
Q 691	29 Aug 44	*Q 708*	9 Dec 44
Q 692	9 Sep 44	*Q 709*	15 Dec 44
Q 693	15 Sep 44	*Q 710*	16 Dec 44
Q 694	21 Sep 44	*Q 711*	20 Dec 44
Q 695	26 Sep 44		

U.S. Army 72-Foot Tug Boats (20)

Vessel	Delivery	Vessel	Delivery
ST 396	13 Aug 43	*ST 406*	25 Jan 44
ST 397	27 Sep 43	*ST 407*	11 Feb 44
ST 398	15 Oct 43	*ST 408*	15 Feb 44
ST 399	19 Oct 43	*ST 409*	6 Mar 44
ST 400	9 Nov 43	*ST 410*	18 Mar 44
ST 401	18 Nov 43	*ST 411*	28 Mar 44
ST 402	24 Nov 43	*ST 412*	31 Mar 44
ST 403	22 Dec 43	*ST 413*	20 Apr 44
ST 404	10 Jan 44	*ST 414*	27 Apr 44
ST 405	1 Jan 44	*ST 415*	3 May 44

U.S. Army 60-Foot Tug Boat (1)

Vessel	Delivery
ST 1930(?)	1944

U.S. Army 50-Foot Salvage Boats (17)

Vessel	Delivery	Vessel	Delivery
J 2266	1945	*J 2275*	19 Feb 45
J 2267	1945	*J 2276*	21 Feb 45
J 2268	1945	*J 2277*	24 Feb 45
J 2269	13 Jan 45	*J 2278*	28 Feb 45
J 2270	27 Jan 45	*J 2279*	28 Mar 45
J 2271	3 Feb 45	*J 2280*	4 Apr 45
J 2272	7 Feb 45	*J 2281*	14 Apr 45
J 2273	13 Feb 45	*J 2282*	20 Apr 45
J 2274	15 Feb 45		

CLYDE W. WOOD

U.S. Army 46-Foot Dry Cargo Barges (25)

Vessel No.	Delivery Date	Vessel No.	Delivery Date
BCS 211	7/42-8/42	*BCS 224*	7/42-8/42
BCS 212	7/42-8/42	*BCS 225*	7/42-8/42
BCS 213	7/42-8/42	*BCS 226*	7/42-8/42
BCS 214	7/42-8/42	*BCS 227*	7/42-8/42
BCS 215	7/42-8/42	*BCS 228*	7/42-8/42
BCS 216	7/42-8/42	*BCS 229*	7/42-8/42

BCS 217	7/42-8/42	*BCS 230*	7/42-8/42
BCS 218	7/42-8/42	*BCS 231*	7/42-8/42
BCS 219	7/42-8/42	*BCS 232*	7/42-8/42
BCS 220	7/42-8/42	*BCS 233*	7/42-8/42
BCS 221	7/42-8/42	*BCS 234*	7/42-8/42
BCS 222	7/42-8/42	*BCS 235*	7/42-8/42
BCS 223	7/42-8/42		

U.S. Army 96-Foot Harbor Tugs (10)

Vessel No.	Delivery Date	Vessel No.	Delivery Date
TP 97	2/44-7/44	*TP 102*	2/44-7/44
TP 98	2/44-7/44	*TP 103*	2/44-7/44
TP 99	2/44-7/44	*TP 104*	2/44-7/44
TP 100	2/44-7/44	*TP 105*	2/44-7/44
TP 101	2/44-7/44	*TP 106*	2/44-7/44

Note: U.S. Army harbor tug *TP 99* named Sgt. *James A. Burzo*, sold as *Margaret Foss* in 1950, later *Baron*, *Royal Baron*

U.S. Army 130-Foot Dry Cargo Barges (11)

Vessel No.	Delivery Date	Vessel No.	Delivery Date
BC 2814	8/44-12/44	*BC 2820*	8/44-12/44
BC 2815	8/44-12/44	*BC 2821*	8/44-12/44
BC 2816	8/44-12/44	*BC 2822*	8/44-12/44
BC 2817	8/44-12/44	*BC 2823*	8/44-12/44
BC 2818	8/44-12/44	*BC 2824*	8/44-12/44
BC 2819	8/44-12/44		

U.S. Coast Guard 38-foot Picket Boats (3)

Vessel No.	Delivery Date	Vessel No.	Delivery Date
CG 4353/38367	12 Mar 42	*CG 4355/38369*	22 Mar 42
CG 4354/38368	18 Mar 42		

Appendix B: Battle Stars Awarded Stockton-built Vessels

Treasury-Bougainville Operation: Treasury Island Landing

Ship	Award Period	Shipbuilder
USS *YMS-96*	27 Oct 43	Stephens Bro. Boat Builders

Marshall Islands Operation: Occupation of Kwajalein and Majuro Atolls

Ship	Award Period	Shipbuilder
USS *YMS-383*	31 Jan-8 Feb 44	Colberg Boat Works
USS *YMS-388*	31 Jan-8 Feb 44	Colberg Boat Works

Marshall Islands Operation: Occupation of Eniwetok Atoll

Ship	Award Period	Shipbuilder
USS *YMS-383* (No BS)	17 Feb-2 Mar 44	Colberg Boat Works

Marianas Operation: Capture and Occupation of Saipan

Ship	Award Period	Shipbuilder
USS *PCS-1402*	15 Jun-28 Jul 44	Colberg Boat Works
USS *PCS-1403*	15 Jun-10 Aug 44	Colberg Boat Works
USS *PCS-1404*	17 Jun-26 Jul 44	Colberg Boat Works
USS *Papaya* (AN-49)	1-10 Aug 44	Pollock-Stockton Shipbuilding

Marianas Operation: Capture and Occupation of Tinian

Ship	Award Period	Shipbuilder
USS *PCS-1402*	24-28 Jul 44	Colberg Boat Works
USS *PCS-1403*	24 Jul-10 Aug 44	Colberg Boat Works
USS *Papaya* (AN-49)	6-15 Aug 44	Pollock-Stockton Shipbuilding

Leyte Operations: Leyte Landings

Ship	Award Period	Shipbuilder
USS *Silverbell* (AN-51)	18 Oct-29 Nov 44	Pollock-Stockton Shipbuilding

Consolidation of Northern Solomons

Ship	Award Period	Shipbuilder
USS *YMS-95*	15 Jun-26 Oct 44	Colberg Boat Works
USS *YMS-97*	15 Jun-26 Oct 44	Colberg Boat Works

Iwo Jima Operation: Assault and Occupation of Iwo Jima

Ship	Award Period	Shipbuilder
USS *ATR-51*	18-25 Feb 45	Colberg Boat Works
USS *ATR-52*	19 Feb-16 Mar 45	Colberg Boat Works
USS *PCS-1403*	19-26 Feb 45	Colberg Boat Works
USS *PCS-1404*	20-26 Feb 45	Colberg Boat Works

Consolidation of Southern Philippines: Mindanao Island Landings

Ship	Award Period	Shipbuilder
USS *Cinnamon* (AN-50)	10-11 Mar 45	Pollock-Stockton Shipbuilding

Okinawa Operation: Assault and Occupation of Okinawa

Ship	Award Period	Shipbuilder
USS *Anchor* (ARS-13)	8-30 June 45	Colberg Boat Works
USS *ATR-51*	1-28 Apr 45	Colberg Boat Works

USS *ATR-53*	30 Jun 45	Colberg Boat Works
USS *YMS-94*	2 Apr-18 Jun 45	Stephens Bro. Boat Builders
USS *YMS-96*	2 Apr-5 May 45	Stephens Bro. Boat Builders
USS *YMS-99*	27-30 Jun 45	Colberg Boat Works
USS *YMS-388*	25 Mar-30 Jun 45	Colberg Boat Works
USS *PCS-1402*	1 Apr-27 Jun 45	Colberg Boat Works
USS *PCS-1403*	1 Apr-30 Jun 45	Colberg Boat Works
USS *PCS-1404*	26 Mar-30 Jun 45	Colberg Boat Works
USS *Snowbell* (AN-52)	26 Mar 45	Pollock-Stockton Shipbuilding
USS *Spicewood* (AN-53)	26 Mar-30 Jun 45	Pollock-Stockton Shipbuilding
USS *Winterberry* (AN-56)	26 Apr-30 Jun 45	Pollock-Stockton Shipbuilding
YH-4 (hospital boat)	24-30 Jun 45	Stephens Bro. Boat Builders

Borneo Operations: Balikpapan Operation

USS YMS-95	15 Jun-9 Jul 45	Colberg Boat Works
USS *YMS-97*	27 Jun-15 Jul 45	Colberg Boat Works

3rd Fleet Operations Against Japan

USS *YMS-388*	5-31 Jul 45	Colberg Boat Works

Special Operations: Minesweeping Operations Pacific: Nagoya (Honshu Area)

USS *YMS-386*	28 Sep-25 Oct 45	Colberg Boat Works
USS *YMS-388*	7-25 Oct 45	Colberg Boat Works

Special Operations: Minesweeping Operations Pacific

USS *YMS-97*	14 Oct 45-2 Mar 46	Colberg Boat Works

Special Operations: Minesweeping Operations Pacific: Fukuoka (Kyushu-Korea Area)

USS *YMS-387*	9-28 Nov 45	Colberg Boat Works
USS *YMS-388* (No BS)	6 Nov-26 Dec 45	Colberg Boat Works

Special Operations: Minesweeping Operations Pacific: Fusan (Kyushu-Korea Area)

USS *YMS-388* (No BS)	13-28 Jan 46	Colberg Boat Works

Special Operations: Minesweeping Operations Pacific: Kobe (Honshu Area)

USS *YMS-386* (No BS)	20 Dec 45-2 Mar 46	Colberg Boat Works
USS *YMS-387* (No BS)	10 Dec 45-2 Mar 46	Colberg Boat Works

Special Operations: Minesweeping Operations Pacific: Yangtze River Approaches

USS *YMS-386* (No BS)	22 Feb-2 Mar 46	Colberg Boat Works
USS *YMS-387* (No BS)	22 Feb-2 Mar 46	Colberg Boat Works
USS *YMS-388* (No BS)	22 Feb-2 Mar 46	Colberg Boat Works

Appendix C: Ships Sunk or Damaged in Support of Okinawa Operation, 26 March-30 April 1945

Name	Type	Date	Cause	Extent
ACHERNAR	AKA 53	2 April	Suicide plane	Major
ADAMS	DM 27	26 March	Suicide plane	Major
AGENOR	ARL 3	5 April	Operational (collision)	Major
ALPINE	APA 92	3 April	Suicide plane	Major
AMMEN	DD 527	21 April	Air Attack	Minor
ARIKARA	ATF 98	11 April	Operational (grounded)	Minor
AUDRAIN	APA 59	6 April	AA fire	Minor
BARNETT	APA 5	26 April	AA fire	Minor
BENNETT	DD 473	7 April	Suicide plane	Major
BENNION	DD 662	28 April	Suicide plane	Minor
BERRIEN	APA 62	11 April	Operational (collision)	Minor
BLACK	DD 666	11 April	Strafing	Minor
BORIE	DD 704	2 April	Operational (collision)	Major
BOWERS	DE 637	16 April	Suicide plane	Major
BOZEMAN VICTORY	XAK	27 April	Suicide boat	Minor
BROWN	DD 546	28 April	Air attack	Minor
BRUSH	DD 745	12 April	Air attack	Minor
BRYANT	DD 665	16 April	Suicide plane	Major
BULL	APD 78	20 April	Operational	Minor
BUSH	DD 529	6 April	Suicide plane	SUNK
BUTLER	DMS 29	28 April	Suicide near miss	Minor
CANADA VICTORY	AKE	27 April	Suicide plane	SUNK
CASSIN YOUNG	DD 793	12 April	Suicide plane	Major
CHILTON	APA 38	3 April	Suicide plane	Minor
C. J. BADGER	DD 657	9 April	Suicide boat	Major
COLHOUN	DD 801	6 April	Suicide plane	SUNK
COLORADO	BB 45	19 April	Operational (powder expl.)	Minor
COMFORT	AH 6	7 April	Air attack	Minor
COMFORT	AH 6	28 April	Suicide plane	Major
CONKLIN	DE 439	12 April	Air attack	Minor
CORREGIDOR	CVE 58	20 April	Storm	Major
COWANESQUE	AO 79	4 April	Storm	Minor
DALY	DD 519	28 April	Suicide near miss	Minor
DAMON M. CUMMINGS	DE 643	12 April	Air attack	Minor
DEFENSE	AM 317	6 April	Suicide plane	Minor
DEVASTATOR	AM 318	6 April	Suicide plane	Minor
DICKERSON	APD 21	3 April	Suicide plane	SUNK
DORSEY	DMS 1	27 March	Suicide plane	Minor

EMMONS	DMS 22	6 April	Suicide plane	SUNK
ENGLAND	DE 635	27 April	Air attack	Minor
ENTERPRISE	CV 6	11 April	Suicide plane (near miss)	Major
ESSEX	CV 9	11 April	Near bomb miss	Minor
FIEBERLING	DE 640	6 April	Suicide plane	Minor
FOREMAN	DE 633	6 April	Air attack	Major
FOREMAN	DE 633	26 March	Suicide plane	Minor
FRANKS	DD 554	2 April	Operational (collision)	Major
GLADIATOR	AM 319	12 April	Suicide plane	Minor
GOODHUE	APA 107	3 April	Suicide plane	Minor
GREGORY	DD 802	7 April	Air Attack	Major
GILMER	APD 11	26 March	Suicide plane	Major
HAGGARD	DD 555	29 April	Suicide plane	Major
HALE	DD 642	11 April	Suicide near miss	Minor
HALE	DD 642	27 April	Operational (collision)	Major
HALLIGAN	DD 584	26 March	Mine	SUNK
HAMBLETON	DMS 20	5 April	Suicide near miss	Minor
HANCOCK	CV 19	6 April	Suicide plane	Major
HANK	DD 702	11 April	Suicide near miss	Minor
HARDING	DMS 28	16 April	Suicide plane	Major
HARRISON	DD 573	6 April	Suicide near miss	Minor
HARRY F. BAUER	DM 26	5 April	Dud torpedo	Minor
HAYNSWORTH	DD 700	7 April	Suicide plane	Major
HAZELWOOD	DD 531	29 April	Suicide plane	Major
HENRICO	APA 45	2 April	Suicide plane	Major
HICKOX	DD 673	26 April	Air attack	Minor
HINSDALE	APA 120	1 April	Suicide plane	Major
HOBBS VICTORY	AKE	6 April	Suicide plane	SUNK
HOBSON	DMS 26	16 April	Suicide plane	Major
HOPPING	APD 51	9 April	Shore battery	Major
HOWORTH	DD 592	6 April	Suicide plane	Major
HUDSON	DD 475	22 April	Air attack	Major
HUTCHINS	DD 476	6 April	Suicide plane	Minor
HUTCHINS	DD 476	27 April	Suicide boat	Major
HYMAN	DD 732	6 April	Suicide plane & torpedo	Major
IDAHO	BB 42	12 April	Suicide plane	Major
INDEFATIGABLE	CV	1 April	Suicide plane	Minor
INDIANAPOLIS	CA 35	31 March	Suicide plane	Major
INTREPID	CV 11	16 April	Suicide plane	Major
ISHERWOOD	DD 520	22 April	Suicide plane	Major
JEFFERS	DMS 27	12 April	Suicide plane	Major
KIDD	DD 661	11 April	Suicide plane	Major
KIMBERLY	DD 521	26 March	Suicide plane	Major
KLINE	APD 120	20 April	Operational (collision)	Minor
KNUDSON	APD 101	26 March	Suicide plane	Minor
LAFFEY	DD 724	15 April	Suicide plane	Major
LCI	82	4 April	Suicide boat	SUNK
LCI	354	9 April	Operational	Minor
LCI	407	16 April	Suicide plane	Minor
LCI	462	2 April	Operational (grounded)	Minor
LCI	558	9 April	Air attack	Minor
LCI(G)	568	2 April	Suicide plane	Minor
LCI	580	2 April	Operational	Minor
LCI	580	8 April	Operational (grounded)	Minor
LCI	580	28 April	Air attack	Major

LCI(G)		588	28 March	Suicide boat	Minor
LCI		754	16 April	Operational	Minor
LCI		765	5 April	Operational	Minor
LCI		816	30 April	Operational (grounded)	Minor
LCI(M)		807	1 April	Operational (mortar expl)	Major
LCS		15	22 April	Suicide plane	SUNK
LCS		33	12 April	Suicide plane	SUNK
LCS		36	11 April	Suicide plane	Major
LCS		37	28 April	Suicide boat	Major
LCS		38	11 April	Operational	Minor
LCS		51	15 April	Suicide plane	Minor
LCS		57	12 April	Operational (grounded)	Minor
LCS		57	12 April	Suicide plane (3)	Major
LCS		88	11 April	Air attack	Major
LCS		116	15 April	Suicide plane	Major
LCT		876	9 April	Suicide plane	SUNK
LEUTZE	DD	481	6 April	Suicide plane	Major
LINDSEY	DM	32	12 April	Suicide plane	Major
LOGAN VICTORY	AKE		6 April	Suicide plane	SUNK
LONGSHAW	DD	559	9 April	Air attack	Minor
LSM		12	16 April	Operational (broached)	SUNK
LSM		28	18 April	Air attack	Major
LSM		188	28 March	Suicide plane	Major
LSM		192	1 April	Operational	Major
LSM		189	15 April	Air Attack	Minor
LSM		193	17 April	Operational (grounded)	Major
LSM		279	12 April	AA fire	Minor
LSM		312	13 April	Operational	Minor
LST		70	5 April	Operational (broaching)	Major
LST		71	5 April	Operational (collision)	Major
LST		78	11 April	Air attack	Major
LST		166	5 April	Operational	Minor
LST		267	19 April	Operational (collision)	Minor
LST		343	9 April	Operational (broaching)	Major
LST		347	6 April	Air attack	SUNK
LST		447	9 April	Suicide plane	SUNK
LST		449	16 April	Shore battery	Minor
LST		554	5 April	Storm	Major
LST		557	10 April	Shore battery	Minor
LST		568	6 April	Operational	Minor
LST		570	6 April	Operational	Minor
LST		599	3 April	Suicide plane	Major
LST		608	5 April	Operational (beaching)	Minor
LST		609	5 April	Operational (beaching)	Minor
LST		612	5 April	Operational (beaching)	Minor
LST		624	5 April	Operational (grounded)	Major
LST		625	5 April	Operational	Minor
LST		658	6 April	Operational	Minor
LST		675	5 April	Operational (grounded)	Major
LST		698	7 April	Operational (beaching)	Major
LST		723	5 April	Operational	Minor
LST		756	5 April	Operational (beaching)	Major

LST		762	5 April	Operational	Minor
LST		782	5 April	Operational	Minor
LST		884	1 April	Suicide plane	Major
LST		890	7 April	Operational (collision)	Minor
LST		929	19 April	Operational (collision)	Minor
MANLOVE	DE	36	10 April	Strafing	Minor
MANNERT L. ABELE	DD	733	12 April	Suicide plane	SUNK
MARYLAND	BB	46	6 April	Bomb	Major
McDERMUT	DD	677	16 April	AA fire	Major
MINOT VICTORY	XAK		14 April	Air attack	Minor
MOBILE	CL	63	19 April	Operational (ex.in turret)	Minor
MORRIS	DD	417	6 April	Suicide plane	Major
MULLANY	DD	528	6 April	Suicide planes (2)	Major
MURRAY	DD	576	27 March	Aerial torpedo	Major
NEVADA	BB	36	26 March	Suicide plane near miss (2)	Major
NEVADA	BB	36	5 April	Shore battery	Minor
NEWCOMB	DD	586	6 April	Suicide planes (2)	Major
NEW MEXICO	BB	40	12 April	Suicide plane	Major
NEW YORK	BB	34	18 April	Suicide plane	Minor
NORTH CAROLINA	BB	55	6 April	AA fire	Minor
NORMAN SCOTT	DD	690	12 April	Air attack	Minor
OAKLAND	CL	95	12 April	Air attack	Minor
O'BRIEN	DD	725	26 March	Suicide plane	Major
PC		462	7 April	Operational	Minor
PC		851	16 April	Operational	Minor
PCM		11	14 April	Operational	Minor
PGM		18	7 April	Mine	SUNK
PINKNEY	APH	2	28 April	Air attack	Major
PORTERFIELD	DD	682	10 April	AA fire	Minor
PORTERFIELD	DD	682	27 March	Suicide plane	Major
PRINGLE	DD	477	16 April	Suicide plane	SUNK
PRITCHETT	DD	561	3 April	Bomb	Major
PURDY	DD	734	12 April	Suicide plane	Major
RALL	DE	304	12 April	Suicide plane	Major
RALPH TALBOT	DD	390	27 April	Air attack	Major
RATHBURNE	APD	25	27 April	Suicide plane	Major
RECRUIT	AM	285	6 April	Suicide plane	Minor
REMEY	DD	688	12 April	Air attack	Minor
R. H. SMITH	DM	23	26 March	Air attack	Minor
RIDDLE	DE	185	12 April	Suicide plane	Major
RODMAN	DMS	21	6 April	Suicide plane	Major
ROOKS	DD	804	6 April	Air attack	Minor
SAMUEL S. MILES	DE	183	12 April	Suicide near miss	Minor
SAN JACINTO	CVL	30	6 April	Bomb near miss	Minor
SC		667	11 April	Operational (grounded)	Major

S. HALL YOUNG	XAK		30 April	Air attack	Minor
SHANNON	DM	25	29 April	Suicide near miss	Minor
SHEA	DM	30	22 April	Air attack	Minor
SIGSBEE	DD	502	13 April	Suicide plane	Major
SKIRMISH	AM	303	26 March	Suicide plane	Minor
SKYLARK	AM	63	26 March	Mines	SUNK
SKIRMISH	AM	303	2 April	Air Attack	Minor
SPEAR	AM	322	19 April	Air attack	Minor
SPROSTON	DD	577	4 April	Bomb near miss	Minor
STANLY	DD	478	12 April	Suicide plane	Major
STARR	AKA	87	9 April	Suicide boat	Minor
STERETT	DD	407	9 April	Suicide plane	Major
STRINGHAM	APD	6	24 April	Operational	Major
SWALLOW	AM	65	22 April	Suicide plane	SUNK
TALUGA	AO	62	16 April	Suicide plane	Major
TAUSSIG	DD	746	6 April	Bomb near miss (2)	Minor
TELFAIR	APA	210	3 April	Suicide plane	Minor
TENNESSEE	BB	43	12 April	Air attack	Major
TERROR	CM	5	30 April	Air attack	Major
THORNTON	AVD	11	5 April	Operational (collision)	SUNK
TOLMAN	DM	28	18 April	Operational (grounded)	Major
TRATHEN	DD	530	14 April	Shore Battery	Major
TWIGGS	DD	591	28 April	Suicide plane	Major
TYRRELL	AKA	80	2 April	Suicide plane	Minor
ULSTER (Brit.)	DD		1 April	Bomb	Major
VAMMEN	DE	644	2 April	Operational	Minor
WADSWORTH	DD	516	28 April	Suicide plane	Minor
WAKE ISLAND	CVE	65	3 April	Suicide near miss	Major
WALTER C. WANN	DE	412	12 April	Air attack	Minor
WESSON	DE	184	7 April	Suicide plane	Major
WEST VIRGINIA	BB	48	1 April	Suicide plane	Minor
WHITEHURST	DE	634	12 April	Suicide plane	Major
WICHITA	CA	45	28 April	Shore battery	Minor
WILSON	DD	408	16 April	Suicide near miss	Minor
WITTER	DE	636	6 April	Suicide planes (2)	Major
WILLIAM J. DITTER	DM	31	30 April	Air attack	Minor
WYANDOT	AKA	92	29 March	Bomb near misses	Minor
YMS		92	8 April	Mine	Major
YMS		96	10 April	Operational (collision)	Major
YMS		103	7 April	Mine	SUNK
YMS		311	6 April	Suicide plane	Minor
YMS		321	6 April	Air attack	Minor
YMS		427	7 April	Shore battery	Minor
ZELLARS	DD	777	12 April	Suicide plane	Major

RECAPITULATION.

*Ships with Major damage	100
*Ships with Minor damage	112
*Ships SUNK	22
Total	234

Los and damaged according to cause:

	Damaged	SUNK
Suicide planes	90	14
Miscellaneous air attack**	47	1
Operational	52	2
Mines	1	4
Shore battery fire	7	0
AA fire	6	0
Suicide boat	6	1
Storm	3	0
Totals	212	22

Lost or damaged by types.***

Type	Damaged	Sunk
BB	10	0
CV	5	0
CA	2	0
CL	2	0
CVL	1	0
CVE	2	0
DD	57	5
DE	16	0
CM	1	0
DM	8	0
DMS	7	1
AVD	0	1
APD	7	1
APA	9	0
AKA	4	0
AO	2	0
AM	7	2
AH	2	0
APH	1	0
ARL	1	0
ATF	1	0
XAK-AKE	3	3
PC	2	0
SC	1	0
PCM	1	1
YMS	5	1
LST	27	2
LSM	7	1
LCT	0	1
LCI	12	1
LCI(M)	1	0
LCS	8	2
Totals	212	22

* NOTE: Some ships were hit more than once for each entry in the list above, but this summary only takes into account the number of entries.

** NOTE: Some of the miscellaneous air attacks may have been by suicide planes.

*** NOTE: Includes British Pacific Fleet ships.

Bibliography / Notes

Bartholomew, Charles A., William I. Milwee Jr., *Mud, Muscle, and Miracles: Marine Salvage in the United States Navy*, 2d ed. Washington, D.C.: Naval History & Heritage Command and Naval Sea Systems Command, 2009.

Bonnett, Wayne. *Build Ships! Wartime Shipbuilding Photographs San Francisco Bay: 1940-1945*. Sausalito, CA: Windgate Press, 1999.

Bruhn, David D. *Eyes of the Fleet: The U.S. Navy's Seaplane Tenders and Patrol Aircraft in World War II*. Berwyn Heights, MD: Heritage Books, 2016.

—*MacArthur and Halsey's "Pacific Island Hoppers" The Forgotten Fleet of World War II*. Berwyn Heights, MD: Heritage Books, 2014.

—*Ready to Haul, Ready to Fight: U.S. Navy, Royal Australian Navy, and British Merchant Navy Cargo Ships in the Pacific in World War II*. Berwyn Heights, MD: Heritage Books, 2021.

Bruhn, David D., Rob Hoole. *Nightraiders: U.S. Navy, Royal Australian Navy, and Royal Netherlands Navy Mine Forces Battling the Japanese in the Pacific in World War II*. Berwyn Heights, MD: Heritage Books, 2018.

Dyer, George Carroll. *The Amphibians Came to Conquer: The Story of Admiral Richmond Kelly Turner*. Washington, D.C.: U.S. Government Printing Office Washington, 1973.

Lott, Arnold S. *Most Dangerous Sea*. Annapolis, MD: U.S. Naval Institute, 1959.

Morison, Samuel Eliot. *Leyte June 1944-January 1945*. Edison, NJ: Castle Books, 1958.

—*New Guinea and the Marianas March 1944-August 1944*. Edison, NJ: Castle Books, 2001.

—*The Liberation of the Philippines: Luzon, Mindanao, the Visayas 1944-1945*. Edison, NJ: Castle Books, 2001.

—*The Two-Ocean War*. Boston: Little, Brown and Company, 1963

Shaw Jr., Henry I., Bernard C. Nalty, and Edwin T. Turnbladh, *History of U.S. Marine Corps Operations in World War II Volume III: Central Pacific Drive*. Washington, DC: Headquarters, U.S. Marine Corps, 1966.

U.S. Government. *Combat Narratives, Solomon Islands Campaign: The Bougainville Landing and the Battle of Empress Bay 27 October-2 November 1943*. Washington, D.C.: Office of Naval Intelligence, 1945.

PREFACE NOTES:

[1] "Commodore Robert Field Stockton: A Legacy of Accomplishment and Controversy 20 August 1795–7 October 1866" (https://www.history.navy.mil/browse-by-topic/people/historical-figures/robert-stockton.html); "Stockton's History" (https://www.visitstockton.org/about-us/stockton-history/): both accessed 22 November 2022.
[2] Wayne Bonnett, *Build Ships! Wartime Shipbuilding Photographs San Francisco Bay: 1940-1945* (Sausalito, CA: Windgate Press, 1999), 39.
[3] Ibid, 40.
[4] Ibid, 39-40.
[5] "California during the Second World War Shipbuilding in Stockton" by Justin Ruhge (http://militarymuseum.org/StocktonShipbuilding.html: accessed 19 November 2022).
[6] Bonnett, *Build Ships!*, 39.
[7] "California during the Second World War Shipbuilding in Stockton" by Justin Ruhge.
[8] Ibid.
[9] Ibid.
[10] Ibid.
[11] David D. Bruhn, *MacArthur and Halsey's "Pacific Island Hoppers" The Forgotten Fleet of World War II* (Berwyn Heights, MD: Heritage Books, 2014), 31-32.
[12] Ibid.
[13] Ibid.
[14] Ibid.
[15] "Battle Honours of RN ships & Naval Air Squadrons" (http://www.royalnavyresearcharchive.org.uk/ESCORT/Battle_hons.htm: accessed 17 August 2020).
[16] Navy and Marine Corps Awards Manual, 1953 (https://www.ibiblio.org/hyperwar/USN/ref/Awards/Awards-IV-17.html#sec2-20: accessed 17 August 2020).
[17] "the ghost fleet" by Ian E. Watts (https://ianewatts.org/army-ships-introduction/: accessed 24 November 2022).
[18] Tim Colton Shipbuilding Records (https://www.shipbuildinghistory.com/: accessed 24 November 2022).

CHAPTER 1 NOTES:

[1] Commanding Officer, USS *PCS-1404*, Action Against Aguijan Island, Marianas, 26 August 1944, report of, 26 August 1944.
[2] Ibid.
[3] USS *PCS-1404* War Diary, March and May 1944.
[4] USS *PCS-1404* War Diary, April 1944.
[5] Ibid.
[6] USS *PCS-1404* War Diary, April and May 1944.
[7] USS *PCS-1404* War Diary, May 1944.

[8] Ibid.
[9] Ibid.
[10] Ibid.
[11] "Transport Doctrine, Amphibious Forces U.S. Pacific Fleet, September 1944" (https://www.ibiblio.org/hyperwar/USN/ref/Transport/transport-41.html#g15: accessed 24 July 2021).
[12] USS *PCS-1404* War Diary, May 1944.
[13] Ibid.
[14] USS *PCS-1404* War Diary, May and June 1944.
[15] USS *PCS-1404* War Diary, June 1944.
[16] Ibid.
[17] Commanding Officer, USS *PCS-1404*, Action Against Aguijan Island, Marianas, 26 August 1944, report of, 26 August 1944.
[18] Commander Escort Division Sixteen War Diary, August 1944.
[19] Commanding Officer, USS *PCS-1404*, Action Against Aguijan Island, Marianas, 26 August 1944, report of, 26 August 1944.
[20] Ibid.
[21] Commanding Officer, USS *PCS-1404*, Action Against Aguijan Island, Marianas, 26 August 1944, report of, 26 August 1944; "Island Surrender" (http://www.uscg83footers.org/island_surrender.htm: accessed 4 November 2022).
[22] Commanding Officer, USS *PCS-1404*, Action Against Aguijan Island, Marianas, 26 August 1944, report of, 26 August 1944.
[23] Ibid.
[24] Ibid.
[25] Ibid.
[26] Ibid.
[27] Commanding Officer, USS *PCS-1404*, Action Against Aguijan Island, Marianas, 26 August 1944, report of, 26 August 1944; USS *PCS-1404* War Diary, August 1944.
[28] Commander Fifth Fleet War Diary, August 1944; Enclosure A of an unidentified report, Aviation History of the Forward Area, Central Pacific and the Marianas Area; Commander in Chief, U.S. Pacific Fleet and Pacific Ocean Areas, Operations in Pacific Ocean Areas – August 1944, 20 January 1945.
[29] "Japanese Surrender of Aguijan Island" by QMCS Larry Richter, USCGR, Ret. (https://web.archive.org/web/20110728145947/http://www.uscg83footers.org/new_page_8.htm); "Island Surrender" (http://www.uscg83footers.org/island_surrender.htm): both accessed 2 November 2022.

CHAPTER 2 NOTES:

[1] *Combat Narratives, Solomon Islands Campaign: The Bougainville Landing and the Battle of Empress Bay 27 October-2 November 1943* (Washington, D.C.: Office of Naval Intelligence, 1945).
[2] Ibid.

[3] Ibid.
[4] Commander Task Group Thirty-One Point One, Report of occupation of the Treasury Islands, 27 October 1943, 10 November 1943.
[5] Ibid.
[6] Ibid.
[7] Ibid.
[8] Southern Force Task Group 31.1 Operation Order No. A16-43, 18 October 1943.
[9] "*YMS-96*" (https://www.navsource.org/archives/11/19096.htm: accessed 27 October 2022)
[10] USS *Renshaw* War Diary, 27 October 1943.
[11] Commander Task Group Thirty-One Point One, Report of occupation of the Treasury Islands, 27 October 1943, 10 November 1943.
[11] Ibid.
[12] "First opposed New Zealand landing since Gallipoli 27 October 1943" (https://nzhistory.govt.nz/first-opposed-new-zealand-landing-since-gallipoli: accessed 16 October 2022); Commander Task Group Thirty-One Point One, Report of occupation of the Treasury Islands, 27 October 1943, 10 November 1943.
[13] Commander Task Group Thirty-One Point One, Report of occupation of the Treasury Islands, 27 October 1943, 10 November 1943.
[14] Ibid.
[15] United States Administration in World War II, CincPac, Motor Torpedo Squadrons, First Draft Narrative Prepared under the General Supervision of the Director of Naval History, p. 38–39; CTG Thirty-One Point One, Report of occupation of the Treasury Islands, 27 October 1943, dated 10 November 1943.

CHAPTER 3 NOTES:

[1] David D. Bruhn, *Eyes of the Fleet: The U.S. Navy's Seaplane Tenders and Patrol Aircraft in World War II* (Berwyn Heights, MD: Heritage Books, 2016), 235.
[2] Ibid, 236.
[3] Ibid, 237.
[4] Commander in Chief, U.S. Pacific Fleet and Pacific Ocean Areas, Operations in the Pacific Ocean Areas – February 1944, 1 June 1944; "Marshall Islands" (https://www.britannica.com/place/Marshall-Islands: accessed 22 December 2020).
[5] Commander in Chief, U.S. Pacific Fleet and Pacific Ocean Areas, Operations in the Pacific Ocean Areas – February 1944, 1 June 1944.
[6] Henry I. Shaw, Jr., Bernard C. Nalty, and Edwin T. Turnbladh, *History of U.S. Marine Corps Operations in World War II Volume III: Central Pacific Drive* (Washington, DC: Headquarters, U.S. Marine Corps, 1966), 125-127; Burton Wright III, *Eastern Mandates*, U.S. Army Center of Military History brochure.
[7] Commander in Chief, U.S. Pacific Fleet and Pacific Ocean Areas, Operations in the Pacific Ocean Areas – February 1944, 1 June 1944.

[8] Commander in Chief, U.S. Pacific Fleet and Pacific Ocean Areas, Operations in the Pacific Ocean Areas – February 1944, 1 June 1944; Commanding Officer, USS *Electra*, Ship's History – forwarding of, 14 October 1945.
[9] USS *Revenge* War Diary, January 1944; Commander in Chief, U.S. Pacific Fleet and Pacific Ocean Areas, Operations in Pacific Ocean Areas – February 1944, 3 June 1944.
[10] Commander in Chief, U.S. Pacific Fleet and Pacific Ocean Areas, Operations in Pacific Ocean Areas – February 1944, 3 June 1944.
[11] USS *Revenge* War Diary, January 1944.
[12] Commander Mine Task Group 52.10.2 War Diary, February 1944.
[13] Ibid.
[14] Ibid.
[15] USS *Revenge* War Diary, February 1944.
[16] Commander Mine Task Group 52.10.2 War Diary, February 1944.
[17] Ibid.
[18] Ibid.
[19] USS *Revenge* War Diary, February 1944.
[20] Commander Mine Task Group 52.10.2 War Diary, February 1944.
[21] Ibid.
[22] Ibid.
[23] Commander, Task Group 52.10, and Commander Mine Task Group 52.10.2 War Diary, February 1944.
[24] Commander in Chief, U.S. Pacific Fleet and Pacific Ocean Areas, Operations in the Pacific Ocean Areas – February 1944, 1 June 1944.
[25] "H-Gram 026: Operations Flintlock, Catchpole, and Hailstone" (https://www.history.navy.mil/content/history/nhhc/about-us/leadership/director/directors-corner/h-grams/h-gram-026.html: accessed 17 December 2020).
[26] Bruhn, *Eyes of the Fleet*, 247-248.

CHAPTER 4 NOTES:

[1] Commander in Chief, United States Fleet, Bulletin No. 17 Battle Experience: Supporting Operations for the Occupation of the Marshall Islands including the Westernmost Atoll, Eniwetok (Washington, DC: February 1944).
[2] Bulletin No. 17 Battle Experience: Supporting Operations for the Occupation of the Marshall Islands including the Westernmost Atoll, Eniwetok; Samuel Eliot Morison, *The Two-Ocean War* (Boston: Little, Brown and Company, 1963 312.
[3] Bulletin No. 17 Battle Experience: Supporting Operations for the Occupation of the Marshall Islands including the Westernmost Atoll, Eniwetok; Commander in Chief, U.S. Pacific Fleet and Pacific Ocean Areas, Operations in Pacific Ocean Areas – February 1944, 3 June 1944.
[4] Commander in Chief, U.S. Pacific Fleet and Pacific Ocean Areas, Operations in Pacific Ocean Areas – February 1944, 3 June 1944.

[5] Commander Task Group Fifty-one Point Fifteen, Report of Minesweeping Operation Eniwetok Atoll, 24 February 1944.
[6] Mine Identification Manual, Navy Department Bureau of Ordnance pamphlet, 1943 (https://maritime.org/doc/mineid/index.php: accessed 31 October 2022).
[7] Commander Task Group Fifty-one Point Fifteen, Report of Minesweeping Operation Eniwetok Atoll, 24 February 1944.
[8] Commander Task Unit Fifty One Point Fifteen Point Two, Eniwetok Action Report – Submission of, 15 March 1944.
[9] Ibid.
[10] Commander Task Unit Fifty One Point Fifteen Point Two, Eniwetok Action Report – Submission of, 15 March 1944; Commander Task Group Fifty-one Point Fifteen, Report of Minesweeping Operation Eniwetok Atoll, 24 February 1944.
[11] Commander Task Unit Fifty One Point Fifteen Point Two, Eniwetok Action Report – Submission of, 15 March 1944.
[12] Ibid.
[13] Bruhn, *Eyes of the Fleet*, 17.
[14] Ibid, 17-18.
[15] Ibid, 19.
[16] Ibid.
[17] Bulletin No. 17 Battle Experience: Supporting Operations for the Occupation of the Marshall Islands including the Westernmost Atoll, Eniwetok.
[18] Ibid.
[19] Ibid.
[20] Ibid.
[21] USS *YMS-262* War Diary, February 1944.
[22] Ibid.
[23] Ibid.
[24] Morison, *The Two-Ocean War*, p. 316.
[25] Ibid, p. 312.
[26] Ibid, p. 314.
[27] Ibid, p. 314-316.
[28] Ibid, p. 316.

CHAPTER 5 NOTES:

[1] George Carroll Dyer, *The Amphibians Came to Conquer: The Story of Admiral Richmond Kelly Turner* (Washington, D.C.: U.S. Government Printing Office Washington, 1973), 861. (https://www.ibiblio.org/hyperwar/USN/ACTC/index.html: accessed 24 July 2021).
[2] CTG 52.2 serial 0226, Enclosure (A).
[3] USS *PCS-1421* War Diary, April 1944.
[4] USS *PCS-1421* War Diary, May 1944.
[5] USS *PCS-1421* War Diary, June 1944; *PCS-1452* War Diary, October 1944.

[6] USS *PCS-1421* War Diary, June 1944.
[7] Ibid.

CHAPTER 6 NOTES:

[1] Commander LST Flotilla 13 War Diary, June 1944.
[2] Ibid.
[3] Ibid.
[4] Ibid.
[5] Ibid.
[6] Ibid.
[7] Ibid.
[8] Samuel Eliot Morison, *New Guinea and the Marianas March 1944-August 1944* (Edison, NJ: Castle Books, 2001), 196-197.
[9] Morison, *New Guinea and the Marianas March 1944-August 1944*, 197; "H-032-1: Operation Forager and the Battle of the Philippine Sea" (https://www.history.navy.mil/content/history/nhhc/about-us/leadership/director/directors-corner/h-grams/h-gram-032/h-032-1.html: accessed 27 July 2021).
[10] Ut supra.
[11] Morison, *New Guinea and the Marianas March 1944-August 1944*, 157,160-161.
[12] Commander Mine Squadron Four War Diary, June and July 1944; Arnold S. Lott, *Most Dangerous Sea* (Annapolis, MD: U.S. Naval Institute, 1959), 175.
[13] Commander Fifth Amphibious Force, U.S. Pacific Fleet War Diary, July 1944.
[14] USS *PCS-1396* War Diary, July 1944.
[15] Ibid.
[16] *PCS-1404* War Diary, June and July 1944.
[17] USS *PCS-461* War Diary, July 1944.
[18] Morison, *New Guinea and the Marianas March 1944-August 1944*, 339-340.
[19] Ibid.

CHAPTER 7 NOTES:

[1] Dyer, *The Amphibians Came to Conquer*, 960.
[2] Dyer, *The Amphibians Came to Conquer*, 958; Commander in Chief, U.S. Pacific Fleet and Pacific Ocean Areas, Operations in Pacific Ocean Areas – July 1944, 22 December 1944.
[3] Commander in Chief, U.S. Pacific Fleet and Pacific Ocean Areas, Operations in Pacific Ocean Areas – July 1944, 22 December 1944.
[4] Commander, Mine Squadron Four War Diary, July 1944; Commander Task Unit 52.13.4, Saipan-Tinian Action Report – Submission of, 4 August 1944; Dyer, *The Amphibians Came to Conquer*, 958.
[5] Commander, Mine Squadron Four War Diary, July 1944.
[6] Commander, Mine Squadron Four War Diary, July 1944; Dyer, *The Amphibians Came to Conquer*, 958.

[7] Dyer, *The Amphibians Came to Conquer*, 867, 950.
[8] Commander, Transport Division Seven, Report of Amphibious Operations – Tinian – July 1944, 1 August 1944.
[9] Ibid.
[10] Ibid.
[11] Ibid.
[12] Ibid.
[13] Commander, LST Flotilla Thirteen War Diary, July 1944.
[14] Ibid.
[15] Commander Task Force Fifty Two, Capture of Tinian – Report of, 24 August 1944.
[16] Ibid.
[17] Ibid.
[18] Ibid.
[19] Ibid.
[20] Ibid.
[21] Ibid.
[22] Dyer, *The Amphibians Came to Conquer*, 956-957.
[23] Ibid, 952.
[24] Ibid, 954.
[25] Ibid, 952-953.
[26] Commander, Task Group 52.8, Action Report on Occupation of Tinian, Mariana Islands, 31 July 1944.
[27] Commanding Officer, USS *Cleveland*, Action Report – bombardment of Japanese installations on Guam, Mariana Islands, 30 July-9 August 1944 (Zone Time minus 10), 11 August 1944.
[28] Commander, Task Group 52.8, Action Report on Occupation of Tinian, Mariana Islands, 31 July 1944.
[29] Commander in Chief, U.S. Pacific Fleet and Pacific Ocean Areas, Operations in Pacific Ocean Areas – July 1944, 22 December 1944.
[30] Ibid.
[31] Ibid.
[32] Commanding Officer, USS *Remey* (DD-688), Action Report of Bombardment of Tinian, 12 August 1944.
[33] Dyer, *The Amphibians Came to Conquer*, 960.

CHAPTER 8 NOTES:

[1] Commander Service Force, U.S Pacific Fleet, History of Service Force – forwarding of, 31 January 1946.
[2] Commanding Officer, USS *Papaya*, Factual History of the USS *Papaya* (AN49), Submission of, 24 October 1945.
[3] Ibid.
[4] Ibid.
[5] Ibid.
[6] Ibid.
[7] Ibid.

[8] Ibid.
[9] Commanding Officer, USS *Papaya*, Factual History of the USS *Papaya* (AN49), Submission of, 24 October 1945; USS *Bowditch* War Diary, August 1944.
[10] Ut supra.
[11] Ut supra.
[12] Ut supra.
[13] USS *Sagittarius* War Diary, August 1944.

CHAPTER 9 NOTES:

[1] Commander in Chief, U.S. Pacific Fleet and Pacific Ocean Areas, Operations in Pacific Ocean Areas – June 1944, 7 November 1944.
[2] Bruhn, *MacArthur and Halsey's "Pacific Island Hoppers,"* 16.
[3] Bruhn, *MacArthur and Halsey's "Pacific Island Hoppers"*; "Northern Solomons: The U.S. Army Campaign of World War II" (https://history.army.mil/brochures/northsol/northsol.htm: accessed 12 November 2022).
[4] "Northern Solomons: The U.S. Army Campaign of World War II."
[5] Commander in Chief, U.S. Pacific Fleet and Pacific Ocean Areas, Operations in Pacific Ocean Areas – June 1944, 7 November 1944.
[6] Ibid.
[7] Ibid.
[8] "William Frederick Halsey, Jr. 30 October 1882 - 16 August 1959" (https://www.history.navy.mil/content/history/nhhc/research/library/research-guides/modern-biographical-files-ndl/modern-bios-h/halsey-william-f.html: accessed 12 November 2022).
[9] Commander Service Squadron, South Pacific Force War Diary, June 1944; Commander in Chief, U.S. Pacific Fleet and Pacific Ocean Areas, Operations in Pacific Ocean Areas – June 1944, 7 November 1944.
[10] "Daniel Edward Barbey 23 December 1889-11 April 1969" (https://www.history.navy.mil/research/library/research-guides/modern-biographical-files-ndl/modern-bios-b/barbey-daniel-edward.html: accessed 13 November 2022).
[11] Commander in Chief, U.S. Pacific Fleet and Pacific Ocean Areas, Operations in Pacific Ocean Areas – June 1944, 7 November 1944.
[12] Commander in Chief, U.S. Pacific Fleet and Pacific Ocean Areas, Operations in Pacific Ocean Areas – July 1944, 22 December 1944.
[13] Commander in Chief, U.S. Pacific Fleet and Pacific Ocean Areas, Operations in Pacific Ocean Areas – August 1944, 20 January 1945.
[14] Commander in Chief, U.S. Pacific Fleet and Pacific Ocean Areas, Operations in Pacific Ocean Areas – September 1944, 7 March 1945.
[15] Ibid.
[16] Commander in Chief, U.S. Pacific Fleet and Pacific Ocean Areas, Operations in Pacific Ocean Areas – October 1944, 31 May 1945.

CHAPTER 10 NOTES:

[1] "USS *ATR-50*" (https://navsource.org/archives/09/40/40050.htm: accessed 13 November 2022)
[2] USS *ATR-50* War Diary, May 1944.
[3] Ibid.
[4] Ibid.
[5] Ibid.
[6] Ibid.
[7] Ibid.
[8] Ibid.
[9] Ibid.
[10] Ibid.
[11] Ibid.
[12] Ibid.
[13] Ibid.
[14] Ibid.
[15] Ibid.
[16] Samuel Eliot Morison, *Leyte June 1944-January 1945* (Edison, NJ: Castle Books, 1958), 92-93.
[17] Ibid, 93.
[18] Ibid, 94-96.
[19] Ibid, 96-97.
[20] Ibid, 98-100.
[21] Morison, *Leyte June 1944-January 1945*, 98-100; *Canberra I* (CA-70), *DANFS*; Commander Cruiser Division Ten, Report of Salvage of USS *Canberra* and USS *Houston*, 30 November 1944.
[22] Morison, *Leyte June 1944-January 1945*, 100.
[23] Ibid.
[24] Commander Cruiser Division Ten, Report of Salvage of USS *Canberra* and USS *Houston*, 30 November 1944.
[25] Ibid.
[26] Ibid.
[27] Ibid.
[28] Ibid.
[29] Ibid.
[30] Ibid.
[31] Ibid.
[32] Ibid.
[33] USS *Canberra* War Diary, October 1944; "*Watch Hill*" (https://www.tugboatinformation.com/tug.cfm?id=7206: accessed 13 November 2022).
[34] Commander Cruiser Division Ten, Report of Salvage of USS *Canberra* and USS *Houston*, 30 November 1944.

CHAPTER 11 NOTES:

[1] USS *Indus* War Diary, October 1944.
[2] Ibid.
[3] USS *Indus* War Diary, October 1944; Commanding Officer, USS *Indus*, Ship's History, 5 October 1945; Morison, *Leyte June 1944-January 1945*, 417.
[4] USS *Indus* War Diary, October 1944.
[5] Commander Task Group 78.7, Action Report – Central Philippines Operation, 10 November 1944; *Arethusa*, *DANFS*.
[6] Commander Task Group 78.7, Action Report – Central Philippines Operation, 10 November 1944.
[7] Ibid.
[8] Bruhn, *Eyes of the Fleet*, 329.
[9] Ibid.
[10] Ibid, 330.
[11] Commander Task Group 79.2, Leyte Operation – Report of, 4 November 1944.
[12] Commander in Chief, U.S. Pacific Fleet and Pacific Ocean Areas, Operations in the Pacific Ocean Areas – October 1944, 31 May 1945.
[13] Ibid.
[14] Bruhn, *Eyes of the Fleet*, 330.
[15] Ibid.
[16] Morison, *Leyte June 1944-January 1945*, 81-82.
[17] Morison, *Leyte June 1944-January 1945*, 83; David D. Bruhn, *Ready to Haul, Ready to Fight: U.S. Navy, Royal Australian Navy, and British Merchant Navy Cargo Ships in the Pacific in World War II* (Berwyn Heights, MD: Heritage Books, 2021), xlii.
[18] Ibid, 84.
[19] "USS *Silverbell* (AN-51)" (https://www.navsource.org/archives/09/18/18051.htm: accessed 23 October 2022).
[20] Commander Service Squadron Four War Diary, June 1945.
[21] Commanding Officer, USS *Silverleaf*, History of USS *Silverleaf* (AN-43), 8 April 1946.
[22] Ibid.
[23] *Teak*, *DANFS*.
[24] USS *Indus*, Ship's History, 5 October 1945.
[25] USS *Indus* War Diary, October 1944.
[26] Ibid.
[27] "Battle of the Philippines" (https://www.history.navy.mil/content/history/nhhc/research/library/online-reading-room/title-list-alphabetically/n/naval-armed-guard-service-in-world-war-ii/battle-of-the-philippines.html: accessed 23 October 2022).
[28] USS *Indus* War Diary, October 1944.
[29] Ibid.
[30] Ibid.
[31] Ibid.

CHAPTER 12 NOTES:

[1] Charles A. Bartholomew and William I. Milwee Jr., *Mud, Muscle, and Miracles: Marine Salvage in the United States Navy*, 2d ed., 2009 (https://www.history.navy.mil/content/dam/nhhc/research/publications/Publication-PDF/Mud%20Muscle%20Miracles_NEW_Web%20Version.pdf: accessed 13 November 2022).
[2] "H-039-4: The First Kaiten Suicide Torpedo Attack, 20 November 1944" (https://www.history.navy.mil/about-us/leadership/director/directors-corner/h-grams/h-gram-039/h-039-4.html: accessed 15 November 2022)
[3] *Mississinewa I* (AO-59), *DANFS*.
[4] Interview of Capt. Philip G. Beek, USNR, by Lieutenant Porter in the Office of Naval Records and Library on 12 December 1944, Personal Interviews, Naval Records and Library, CNO.
[5] Ibid.
[6] Ibid.
[7] "H-039-4: The First Kaiten Suicide Torpedo Attack, 20 November 1944."
[8] Ibid.
[9] Ibid.
[10] Ibid.
[11] Ibid.
[12] Ibid.
[13] Commanding Officer, USS *ATR-51*, Report on Fire, 20 November 1944.
[14] Ibid.
[15] Ibid.
[16] Interview of Capt. Philip G. Beek, USNR, by Lieutenant Porter in the Office of Naval Records and Library on 12 December 1944, Personal Interviews, Naval Records and Library, CNO.
[17] Ibid.
[18] Ibid.
[19] Ibid.

CHAPTER 13 NOTES:

[1] David D. Bruhn and Rob Hoole, *Nightraiders: U.S. Navy, Royal Australian Navy, and Royal Netherlands Navy Mine Forces Battling the Japanese in the Pacific in World War II* (Berwyn Heights, MD: Heritage Books, 2018), 241.
[2] Ibid.
[3] Ibid, 241-242.
[4] Commander in Chief, U.S. Pacific Fleet and Pacific Ocean Areas, Operations in Pacific Ocean Areas – February 1945, 27 August 1945.
[5] Ibid.
[6] USS *PCE-877* War Diary, February 1945.
[7] Ibid.
[8] USS *PCS-1452* War Diary, February 1945.
[9] Ibid.
[10] Ibid.
[11] Ibid.

[12] Commanding Officer, USS *PCS-1455*, Action Report, Invasion of Iwo Jima, 19 February 1945, 19 March 1945.
[13] USS *PCS-1403* War Diary, February 1945.
[14] Ibid.
[15] Ibid.
[16] Ibid.
[17] Ibid.
[18] Ibid.
[19] Ibid.
[20] USS *PCS-1403* War Diary, February 1945; USS *Gear* (ARS-34) War Diary, March 1945.
[21] Bruhn, *Eyes of the Fleet*, 359.
[22] "In the 1940s, the U.S. Navy Launched Planes From Trapezes" by Steve Weintz (https://medium.com/war-is-boring/how-the-navy-launched-planes-from-a-trapeze-9e1032cac130: accessed 18 November 2022).
[23] Ibid.
[24] Ibid.
[25] Ibid.
[26] Commander in Chief, U.S. Pacific Fleet and Pacific Ocean Areas, Operations in Pacific Ocean Areas – February 1945, 27 August 1945; CincPac War Diary, 27 January-28 February 1945; USS ATR-52 War Diary, February 1945; Bartholomew and Milwee Jr., *Mud, Muscle, and Miracles: Marine Salvage in the United States Navy*, 179-180.
[27] Bartholomew and Milwee Jr., *Mud, Muscle, and Miracles: Marine Salvage in the United States Navy*, 54.
[28] Ibid, 58.
[29] Ibid, 91, 149, 154.
[30] Ibid, 179-180.
[31] Ibid.
[32] USS *ATR-52* War Diary, February and March 1945.

CHAPTER 14 NOTES:

[1] Samuel Eliot Morison, *The Liberation of the Philippines Luzon, Mindanao, the Visayas 1944-1945* (Edison, NJ: Castle Books, 2001), 217.
[2] Commander in Chief, U.S. Pacific Fleet and Pacific Ocean Areas, Operations in Pacific Ocean Areas – March 1945, 31 August 1945.
[3] "HMAS *Warrego* (II)" (https://www.navy.gov.au/hmas-warrego-ii: accessed 24 October 2022); Chapter "History of the Australian Hydrographic Service" by Kevin Slade, in *100 Years of the Royal Australian Navy*, edited by Charles Oldham (Bondi Junction, NSW: Faircount Media Group, 2011).
[4] "USS *Cinnamon* (AN-50)" (https://navsource.org/archives/09/18/18050.htm: accessed 24 October 2022).
[5] Commander in Chief, U.S. Pacific Fleet and Pacific Ocean Areas, Operations in Pacific Ocean Areas – March 1945, 31 August 1945.

[6] Commander Task Group 78.1 (Commander Amphibious Group Six), Amphibious Attack on Zamboanga, Mindanao – Report of, 26 March 1945.
[7] Commander in Chief, U.S. Pacific Fleet and Pacific Ocean Areas, Operations in Pacific Ocean Areas – March 1945, 31 August 1945.
[8] Command Officer, USS *Boise*, Report of Operations in Support of Amphibious Landings at Zamboanga, Mindanao, Philippine Islands, 8-18 March 1945, inclusive, 18 March 1945.
[9] Ibid.
[10] Commander in Chief, U.S. Pacific Fleet and Pacific Ocean Areas, Operations in Pacific Ocean Areas – March 1945, 31 August 1945.
[11] Ibid.
[12] Ibid.
[13] Ibid.
[14] Ibid.

CHAPTER 15 NOTES:

[1] Commander Task Group 32.2 (52.2), Report of Capture of Okinawa Gunto – Phases One and Two, 23 July 1945.
[2] Commander in Chief, U.S. Pacific Fleet and Pacific Ocean Areas, Operations in the Pacific Ocean Areas – April 1945, 16 October 1945.
[3] Ibid.
[4] Ibid.
[5] Commander Task Force 52, Action Report – Operations against Okinawa Gunto Including the Capture of Kerama Retto and the Eastern Islands of Okinawa – March 21 to and Including April 20, 1945, 1 May 1945.
[6] Commander in Chief, U.S. Pacific Fleet and Pacific Ocean Areas, Operations in the Pacific Ocean Areas – April 1945, 16 October 1945; Commander Task Group 32.2 (52.2), Report of Capture of Okinawa Gunto – Phases One and Two, 23 July 1945.
[7] USS *Adams* (DM-27) War Diary, March 1945.
[8] Commander, Mine Squadron Four, Action Report of Task Group 52.4 – Okinawa Operation – 4 to 31 March 1945 – Forwarding of, 20 April 1945.
[9] "The Radar Picket Line" (The Radar Picket Line - American Amphibious Gunboats in World War II: A History of LCI and LCS(L) Ships in the Pacific (erenow.net): accessed 16 November 2021).
[10] Ibid.
[11] Commanding Officer, USS *YMS-311*, General Action Report "Report of Capture of Okinawa Gunto, Phases I and II," Forwarding of, 19 July 1945.
[12] Commander Minecraft, U.S. Pacific Fleet, General Action Report "Report of Capture of Okinawa Gunto, Phases I and II," Forwarding of, 9 August 1945.
[13] Commanding Officer, USS *YMS-311*, General Action Report "Report of Capture of Okinawa Gunto, Phases I and II," Forwarding of, 19 July 1945; Commanding Officer, USS *Bowers* (DE-637), Action Report of the USS *Bowers* (DE-637), 1-24 April 1945, 15 May 1945.

[14] Commanding Officer, USS *YMS-311*, General Action Report "Report of Capture of Okinawa Gunto, Phases I and II," Forwarding of, 19 July 1945.
[15] "The Evolution of Ship Naming in the U.S. Navy" (https://www.history.navy.mil/content/history/nhhc/browse-by-topic/heritage/customs-and-traditions0/ship-naming/the-evolution-of-ship-naming-in-the-u-s--navy.html: accessed 11 November 2021).
[16] Commanding Officer, USS *YMS-311*, General Action Report "Report of Capture of Okinawa Gunto, Phases I and II," Forwarding of, 19 July 1945.
[17] Ibid.
[18] Ibid.
[19] Ibid.
[20] Commanding Officer, USS *YMS-311*, Action Report, 10 June 1945.
[21] USS *YMS-311* Crew Muster, Report of Changes, for the month ending 31 May 1945, 16 June 1945; Commanding Officer, USS *YMS-311*, General Action Report "Report of Capture of Okinawa Gunto, Phases I and II," Forwarding of, 19 July 1945.
[22] Commanding Officer, USS *YMS-311*, General Action Report "Report of Capture of Okinawa Gunto, Phases I and II," Forwarding of, 19 July 1945.
[23] Commander in Chief, U.S. Pacific Fleet and Pacific Ocean Areas, Operations in the Pacific Ocean Areas – April 1945, 16 October 1945.
[24] Ibid.
[25] "The Most Difficult Antiaircraft Problem Yet Faced By the Fleet: U.S. Navy vs. Kamikazes at Okinawa" (https://www.history.navy.mil/browse-by-topic/wars-conflicts-and-operations/world-war-ii/1945/battle-of-okinawa/antiaircraft-problem.html: accessed 19 November 2021).
[26] Ibid.
[27] Bruhn, *Eyes of the Fleet*, 361-362.

CHAPTER 16 NOTES:

[1] Commander Amphibious Forces, U.S. Pacific Fleet, General action report, Capture of Okinawa Gunto, Phases I and II, 17 February 1945 to 17 May 1945, Submission of, 25 July 1945.
[2] Ibid.
[3] Ibid.
[4] Ibid.
[5] Commander Amphibious Group Four, Pacific Fleet, Report of Participation in the Capture of Okinawa Gunto – Phases I and II, 20 July 1945. This document reported *PCS-1390* being assigned to Control Unit Able, but other references indicate she was replaced by *PCS-1460*.
[6] Commanding Officer, USS *PCS-1452*, Action Report for Invasion of Okinawa Jima, 9 April 1945.
[7] Ibid.
[8] Ibid.
[9] USS *PCS-1403* War Diary, April 1945.
[10] Ibid.
[11] Ibid.

[12] Commander Amphibious Forces, U.S. Pacific Fleet, General action report, Capture of Okinawa Gunto, Phases I and II, 17 February 1945 to 17 May 1945, Submission of, 25 July 1945.
[13] Commander Amphibious Group Seven (CTG 51.1), Action Report – Capture of Okinawa Gunto, Phases 1 and 2, 26 May 1945.
[14] Ibid.
[15] Ibid.
[16] Ibid.
[17] Ibid.
[18] Commander Service Squadron Twelve, Action Report of Attack by Enemy Suicide Plane, 8 June 1945.
[19] *Dutton*, *DANFS*.
[20] Commanding Officer, USS *Dutton* (AGS-8), Action Report of Attack by Enemy Suicide Planes, 28 May 1945.
[21] Ibid.
[22] Ibid.
[23] Ibid.
[24] Commanding Officer, USS *Dutton* (AGS-8), Action Report of Attack by Enemy Suicide Planes, 28 May 1945; Commanding Officer, USS *Dutton* (AGS-8), Damage Resulting from Ship being Crashed by Enemy Suicide Plane, 28 May 1945.
[25] Commanding Officer, USS *Dutton* (AGS-8), Action Report of Attack by Enemy Suicide Planes, 28 May 1945.
[26] Ibid.
[27] Ibid.
[28] Commanding Officer, USS *Dutton* (AGS-8), Action Report of Attack by Enemy Suicide Planes, 28 May 1945; *Dutton*, *DANFS*.
[29] Commander Amphibious Group Seven (CTG 51.1), Action Report – Capture of Okinawa Gunto, Phases 1 and 2, 26 May 1945.
[30] Ibid.

CHAPTER 17 NOTES:

[1] USS *Salem* War Diary, March 1945.
[2] Ibid.
[3] Ibid.
[4] The Former Executive Officer, Ship's History of USS *Snowbell* (AN-52), 29 January 1946.
[5] Ibid.
[6] Ibid.
[7] Ibid.
[8] Ibid.
[9] The Former Executive Officer, Ship's History of USS *Snowbell* (AN-52), 29 January 1946; USS *Salem* War Diary, March 1945.
[10] USS *Salem* War Diary, March 1945.
[11] The Former Executive Officer, Ship's History of USS *Snowbell* (AN-52), 29 January 1946.

[12] Ibid.
[13] Ibid.
[14] Ibid.
[15] Ibid.
[16] USS *Anchor* War Diary, June 1945.
[17] Ibid.
[18] Ibid.
[19] Ibid.
[20] Ibid.
[21] Commanding Officer, USS *LCS(L)(3)-116*, Action Report, operations in vicinity of Okinawa, 1 April to 16 April, 1945, 18 April 1945.
[22] USS *ATR-53* War Diary, July 1945.

CHAPTER 18 NOTES:

[1] Bruhn and Hoole, *Nightraiders*, 297.
[2] Commander Task Group 78.2, Action Report – Balikpapan-Manggar-Borneo June 15 – July 6 1945, 1 August 1945.
[3] Commander Task Group 78.2, Action Report – Balikpapan-Manggar-Borneo June 15 – July 6 1945, 1 August 1945; Commander, Task Unit 78.2.9, Action Report, Minesweeping Unit Balikpapan, Borneo, NEI, 11 June to 1 July 1945, 11 July 1945.
[4] Commander, Task Unit 78.2.9, Action Report, Minesweeping Unit Balikpapan, Borneo, NEI, 11 June to 1 July 1945, 11 July 1945.
[5] Ibid.
[6] Ibid.
[7] Ibid.
[8] Ibid.
[9] Ibid.
[10] Commander, Task Unit 78.2.9, Action Report, Minesweeping Unit Balikpapan, Borneo, NEI, 11 June to 1 July 1945, 11 July 1945; Commanding Officer, USS *YMS-95*, Action Report, USS YMS 95, Balikpapan operations, from 11 June, to 2 July, 1945, 2 July 1945.
[11] Commander, Task Unit 78.2.9, Action Report, Minesweeping Unit Balikpapan, Borneo, NEI, 11 June to 1 July 1945, 11 July 1945.
[12] Commander, Task Unit 78.2.9, Action Report, Minesweeping Unit Balikpapan, Borneo, NEI, 11 June to 1 July 1945, 11 July 1945; Lott, *Most Dangerous Sea*, 162.
[13] Commander, Task Unit 78.2.9, Action Report, Minesweeping Unit Balikpapan, Borneo, NEI, 11 June to 1 July 1945, 11 July 1945.
[14] Ibid.
[15] Commanding Officer, USS *YMS-95*, Action Report, USS *YMS 95*, Balikpapan operations, from 11 June, to 2 July, 1945, 2 July 1945.
[16] Commander, Task Unit 78.2.9, Action Report, Minesweeping Unit Balikpapan, Borneo, NEI, 11 June to 1 July 1945, 11 July 1945.
[17] Ibid.
[18] Ibid.

[19] Ibid.
[20] Ibid.
[21] Commander, Task Unit 78.2.9, Action Report, Minesweeping Unit, Balikpapan, Borneo, NEI, 11 June to 1 July 1945, 11 July 1945; Lott, *Most Dangerous Sea*, 161-163.
[22] Commander, Task Unit 78.2.9, Action Report, Minesweeping Unit Balikpapan, Borneo, NEI, 11 June to 1 July 1945.
[23] Lott, *Most Dangerous Sea*, 163.
[24] Bruhn and Hoole, *Nightraiders*, 297.
[25] Ibid.
[26] Ibid.

POSTSCRIPT NOTES:

[1] Bruhn, *Eyes of the Fleet*, 397.
[2] Ibid, 393.
[3] Ibid, 395.
[4] Ibid, 391.
[5] Lott, *Most Dangerous Sea*, 246; Commanding Officer, USS *Requisite* (AM-109), Action Report, Minesweeping of China Sea Operation Area, 1 August 1945.

Index

About the Author

Commander David D. Bruhn, U.S. Navy (Retired) served twenty-two years on active duty and two in the Naval Reserve, as both an enlisted man and as an officer, between 1977 and 2001.

He is a graduate of California State University, Chico, and has Masters Degrees from the U.S. Naval Postgraduate School and U.S. Naval War College.

During his career, Bruhn served aboard six ships including command of the mine countermeasures ships USS *Gladiator* (MCM-11) and USS *Dextrous* (MCM-13) in the Persian Gulf. Ashore, he did two three-year tours in the Pentagon. During the first one, he was assigned to Secretary of the Navy and Chief of Naval Operation staffs as a budget analyst and resources planner. His final assignment was to the Secretary of Defense staff as executive assistant to a senior (SES 4) executive at the Ballistic Missile Defense Organization in Washington, D.C.

Following military service, he was a high school teacher and track coach for ten years, and remains an avid Track & Field fan. He lives in northern California with his wife Nancy and has two grown sons, David and Michael.

A prolific writer, Bruhn has authored twenty-five books on naval history; one on shipboard engineering; and two related to sports, *Toe the Mark* and *Stride Out* about competitive running in the 1970s.

www.ingramcontent.com/pod-product-compliance
Lightning Source LLC
LaVergne TN
LVHW020537100826
845148LV00010B/1496

9780788429125